Proton Structure

William Stubbs

In memory of my eighth-grade science teacher, Ms. Bradshaw

Contents

List of Tables

List of Figures

Preface

Some of the hardest things to see are the things right in front of you. You look at them over, and over again, day in and day out, but you never really see what they are. You only see what you have been told they are, by others who see what they have been told. It is very safe to see what everyone else sees.

In this book, I look at data collected over the past 50 years to see what it reveals about the structure of the proton. This undertaking has been done countless times before now, by many people a lot smarter than I am. However, just as looking at photographs of a place is no substitute for actually being there; I sought to make the journey through the information to personally experience its teachings, rather than rely on the accounts of others.

My focus was on understanding and explaining the dynamics of the forest instead of dwelling on the intricacies of the many trees forming it. In that spirit, I present the high-level concepts that my assessments are based on, but avoided discussing the rigorous mathematics used to formulate those concepts. I do provide references to the derivations of these concepts in the endnotes.

All of the data that I used are in the appendices of the book, along with formal references to them. I have done this because of past frustrations I experienced when trying to recreate information presented in graphs and charts of other works. Although data may be available, it is not always easily accessible.

One of the most difficult things to do is to question what everyone already knows to be. Not because of the resistance you will encounter from those who already know the answers, but because you likely are one of those who already knows the answers. A known answer is a very hard thing to question.

William L. Stubbs
Port St. Lucie, Florida
March 2015

Proton Structure

1
Introduction

If we isolate an atom of a radioactive isotope, say potassium-52, and watch it continuously, in time it will emit a negative beta particle. A negative beta particle is a particle having a mass and a charge equal to those of the electron. Since the neutral potassium atom has 19 electrons in it, it seems plausible that the atom could eject one of these electrons, causing the observed event. However, careful examination of the resulting atom reveals that, instead of the 18 electrons expected to surround the nucleus because of the one emitted; there are actually 20 electrons now orbiting the nucleus.

Although the nucleus appears to contain the same number of nucleons after the beta particle emission as before, it seems to have gained a proton and lost a neutron, because its charge is now 20. Since, with a nuclear charge of 20, the atom is calcium; because of the beta emission, the potassium atom has become an atom of calcium. It is the isotope calcium-52, another radioactive isotope.

Because the emission of the beta particle caused the atom to change from potassium to calcium, the emitted particle does not appear to have come from the atom's orbital electrons. In order to change the type of element an atom is, a change must occur to its nucleus. The beta particle emission appears to have changed the nucleus of the potassium-52 atom into that of a calcium-52 atom. The emitted beta particle appears to have originated from the nucleus of the potassium-52 atom.

If we continue to watch the newly formed calcium atom, it will eventually emit another negative beta particle. The charge of the nucleus once again changes, now increasing from 20 to 21, making the atom an isotope of scandium. Again, the beta emission does not significantly alter the mass of the atom; its mass remains 52, the mass of the original potassium atom. This means that, what originally began as a potassium-52 atom; appears to have emitted two negative beta particles. In the process, two of its neutrons appear to have become protons, making it appear as if the nucleus ejected the beta particles. This resulted in the original potassium-52 atom becoming scandium-52, which is another radioactive isotope.

Over time, in addition to the two negative beta emissions discussed above, our atom will emit three more beta particles. The scandium-52 will emit a particle to become titanium-52, and the titanium-52 will emit a particle to become vanadium-52, both radioactive. Finally, the vanadium-52 will become chromium-52, which is stable. In each case, the atom emits a negative beta particle that causes the charge of the nucleus to increase by one, converting the atom into its neighboring element.

Figure 1.1 shows that the emission of five negative beta particles from the original potassium-52 atom ultimately formed the stable chromium-52 atom. Since each emission caused the emitting isotope to change, the beta particles appear to come from within the nucleus. The emission of negative beta particles appears to cause neutrons to turn into protons. A proposed explanation of this phenomenon at one time was that neutrons are complex particles consisting of a proton and an electron.[1] When negative beta decay occurs, one of the neutrons loses its electron, and becomes a proton. In the case of our potassium-52 atom, ultimately five of its neutrons would lose their electrons, causing five new protons to form. This transforms the potassium into chromium.

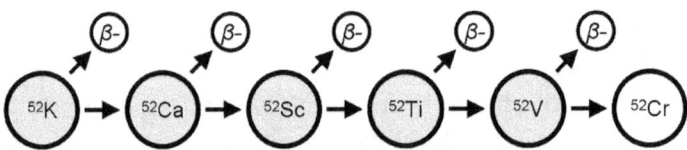

Figure 1.1: *The decay sequence for the potassium-52 atom. It emits five negative beta particles before becoming the stable chromium-52 atom.*[2]

Similarly, if we isolate and watch an atom of the radioactive isotope nickel-52, it, too, will eventually emit a beta particle. However, the beta particle it emits has a positive unit charge instead of the negative charge we saw in the potassium-52 emission. It still carries the mass of an electron, but its positive unit charge makes it appear identical to the positron, the antiparticle of the electron. Nickel-52 emits a positive beta particle. In this case, when the atom emits the particle, the mass of the resulting atom remains 52; however, the charge of the nucleus decreases by one from 28 to 27. This makes the resulting atom an isotope of cobalt. The positive beta emission by the nickel-52 atom causes it to become a cobalt-52 atom, another radioactive isotope.

Since there are only electrons surrounding the nucleus of the atom, this positive beta particle could not have come from the atom's electron cloud. Because the emission of the positive beta particle changed the charge of the nucleus, as with the negative beta particle, it appears to be emanating from the nucleus. At first glance, the resulting atom seems to suggest that the emission of a positive beta converts a proton into a neutron during the process. However, as far as we know, neutrons are more massive than protons.[3] If the proton lost the beta particle during the emission, while its charge would become neutral like a neutron, its mass would be significantly less than that of a neutron.

Watch a while longer and our cobalt-52 atom will give off another positive beta particle. Again, the mass of the atom stays at 52, but the charge of its nucleus decreases from 27 to 26, converting the atom to an isotope of iron. The new atom is iron-52, which is also a radioactive isotope. Now our original nickel-52 atom has emitted two positive beta particles to become an iron-52 atom, even though there does not appear to be a source of positive beta particles within the atom from which the emissions can occur.

Our iron-52 atom will emit two additional positive beta particles over time, ultimately becoming the stable isotope chromium-52. The iron-52 atom will emit a positive beta particle to become the radioactive isotope manganese-52; and the manganese-52 atom will emit a positive beta particle to become an atom of the stable isotope chromium-52. In each case, a positive beta particle is materializing from somewhere within the nucleus of the atom and being ejected.

Figure 1.2 shows that, what is originally a radioactive nickel-52 atom emits four positive beta particles to become stable chromium-52. Like the negative beta particles from the potassium-52 atom's transitions, the release of each positive beta particle by the atom changed the isotope. Therefore, the positive beta particles also appear to be coming from within the nucleus. However, as previously discussed, there does not appear to be a source of positive beta particles or positrons within the atom.

At one time, some thought that a proton might be a positron bound to a neutron, similar to the suggestion of a neutron made of an electron bound to a proton.[4] However, neutrons are already heavier than protons. Adding another particle to it to form a proton would make it even heavier. Nonetheless, protons do appear to be converting into neutrons when the atoms emit positive beta particles. With no other explanation handy, assuming that a proton is a neutron and a positron provides a mechanism for the emission.

The neutron-positron proton may not be that farfetched. Protons and neutrons appear to lose some of their mass when they are part of a nucleus. If the protons and neutrons in the nucleus are less massive than their free counterparts are; then the nuclear protons could be undersized neutrons with positrons attached to them. This would create a source of positrons in the nucleus. Then, when a nucleus emits a positron, one of its protons would become a neutron. This is a convenient way to view this process, for now.

If the nickel-52 atom contains the neutron-positron protons; then when it emits four positive beta particles, four of the protons in its nucleus would convert into neutrons. This would convert the nickel atom into a chromium atom. We base this solely on the assumption that the nucleus contains positive beta particles. The fact that these nuclei emit these particles makes a strong case for the nuclei containing them.

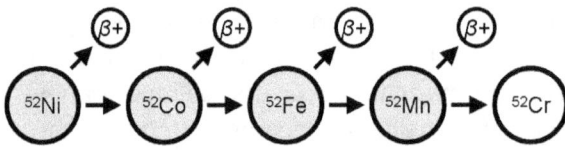

Figure 1.2: *The decay sequence for the nickel-52 atom. It emits four positive beta particles before becoming the stable chromium-52 atom.*[5]

The beta decay examples above suggest that the nucleus of an atom contains a collection of negative and positive beta particles. The potassium-52 nucleus appears to contain at least five negative beta particles; the calcium-52 at least four negative beta particles; the scandium-52 at least three negative beta particles; titanium-52 at least two negative beta particles; and vanadium-52, at least one negative beta. In all likelihood, if the nuclei of these isotopes contain negative beta particles, they contain more negative beta particles than were emitted during the decays, which means the stable chromium-52 nucleus probably also contains negative beta particles.

Similarly, the nickel-52 nucleus appears to contain at least four positive beta particles; the cobalt-52 nucleus at least three positive beta particles; the iron-52 at least two positive beta particles; and the manganese-52, at least one positive beta particle. Again, if these nuclei contain positive beta particles, they probably contain more positive beta particles than were emitted during the decays. This means that the stable chromium-52 nucleus probably also contains some positive beta particles. The two beta decay chains discussed seem to suggest that a nucleus contains both positive and negative beta particles. The decay of the copper-64 atom (Figure 1.3) makes an even stronger case for this.

If we watch a collection of copper-64 atoms, we will see that some of the copper atoms appear to emit negative beta particles becoming zinc-64, while others appear to emit positive beta particles to become nickel-64. It appears copper-64 can emit either a negative beta particle or a positive beta particle from its nucleus. As far as we know, there are not two kinds of copper-64 nuclei. There is not one kind of copper-64 that becomes zinc-64 and another kind that becomes nickel-64. All of the copper-64 nuclei are the same. Therefore, it appears that the copper-64 nucleus has both negative beta particles and positive beta particles in it.

Figure 1.3: *The two decay paths for the copper-64 atom. It can emit either a negative beta particle to become zinc-64, or a positive beta particle to become nickel-64.*[6]

Since there is no reason to believe that the basic components of the copper nucleus are different from those of any other nucleus, it seems reasonable to assume that all nuclei may contain both negative and positive beta particles. The negative beta particle has the same mass and charge as an electron and the positive beta particle has the same mass and charge as a positron. Negative beta particles appear to be electrons and positive beta particles, positrons. There-fore, based on the beta decay of nuclei, electrons and positrons appear to reside within the nucleus. Since the nucleus is made of only protons and neutrons, if there are electrons and positrons in it, they must reside within protons and neutrons. In other words, it appears protons and neutrons are made of electrons and positrons.

If protons and neutrons are made of electrons and positrons, then electrons and positrons appear to play a fundamental role in the structure of matter. Since the only components of the atom other than protons and neutrons are electrons, if electrons and positrons make up the protons and neutrons, then the entire atom becomes a collection of electrons and positrons.

There is actually experimental data that could be showing that protons and neutrons are made of electrons and positrons. The prevailing interpretations of this data indicate otherwise. However, it is possible to make a strong case for this position. The following discussion attempts to make that case.

2
Electron Scattering

In 1967, a set of experiments done at the Stanford Linear Accelerator Center (SLAC) fired high-energy electrons into proton targets.[7] These experiments allowed physicists to probe the interior of the proton by analyzing the scattering the electrons experienced from their collisions with the proton. These experiments were akin to those performed by Ernest Rutherford from 1906 through 1911 that led to the discovery of the atomic nucleus.[8]

The experiments done at SLAC used electrons with energies up to about 20 GeV. Similar experiments, done at the High Energy Physics Laboratory (HEPL) at Stanford University in the 1950s, used electrons with energies of up to 0.5 GeV.[9] At 0.5 GeV, the wavelength of the electron, which is its size going into the collision, is about 2.5×10^{-15} m. This is about the size of the proton. Therefore, these proton-electron collisions are elastic. The two colliding particles just bounce off each other, retaining their identities after the collision. These types of collisions can only reveal the size of the proton. The 1950s experiments determined that the proton is a complex particle, rather than the simple point particle previously thought, with a diameter of about 1.5×10^{-15} m.

At 20 GeV, the electron's wavelength is in the 0.06×10^{-15} m range, one twenty-fifth the proton diameter. Now, instead of bouncing off the proton, the electrons can penetrate it and interact with components inside it, if there are any. Turns out there are internal components within the proton and the electrons scatter off them. When this happens, instead of retaining its identity, the proton may shatter into a collection of smaller particles, which makes the scattering inelastic. The electron's inelastic scattering becomes the sum of all its elastic scatters off the internal components of the proton. This is deep inelastic scattering (DIS).[10] The energies and scattering angles of the electrons after the collisions reveal information about the components within the proton.

There are three fundamental variables in proton DIS physics.[11] The first is the invariant momentum transferred by the electron to the proton during the collision, designated Q^2. Next is the energy transferred from the electron to the proton during the collision, ν. Finally, there is a scaling factor, x, which is also the fraction of the proton's momentum carried by its component the electron hits. We can calculate these by knowing the initial electron energy, E_0, the electron energy after the collision, E', and the scattering angle, θ, of the electron. The momentum transfer Q^2 is

$$Q^2 = 4E_0 E' \sin^2\left(\frac{\theta}{2}\right), \qquad (2.1)$$

the energy transferred by the electron, ν is

$$\nu = E_0 - E', \qquad (2.2)$$

and the scaling factor x becomes

$$x = \frac{Q^2}{2M_p \nu}, \qquad (2.3)$$

where M_p is the mass of the proton, 0.938 GeV.

As an example, in an actual deep inelastic scattering run at SLAC, the initial energy of the electron was 12.518 GeV. It scattered off the proton at an angle of 34.009°, having 2.75 GeV of energy after the collision.[12] For this event, the momentum transferred to the proton was $Q^2 = 11.777$ GeV², the energy transferred was $\nu = 9.786$ GeV, and the scaling factor was $x = 0.642$.

These momentum transfer, energy transfer, and scaling factor go into developing structure functions that describe the internals of the target. High-energy scattering physics reveals that two structure functions, F_1 and F_2, model the proton. The F_1 function models charge distribution, and the F_2 function, momentum distribution. J.D. Bjorken predicted that if the proton contains particles, then there is an energy region where these structure functions do not change as the momentum transferred by the electrons to the protons during collisions increases, a phenomenon called Bjorken scaling.[13]

Think of the structure functions as the picture within the proton. The wavelengths of low energy electrons are large and are like using a low resolution to view the picture. As the energy of the electrons used to see the proton increases, their wavelengths get smaller and the resolution of the picture gets better. However, at some point, the picture is as sharp as it can get. Increasing the energy of the electrons used for viewing, which amounts to increasing the resolution, does not improve the sharpness of the picture. This is what Bjorken expected to happen if there were things to see inside the proton. Increasing the energy of the electrons up to some point would produce a sharper image of the inside of the proton, but beyond that point, there would be no change.

The results of the SLAC experiments showed that for momentum transfers greater than 4 GeV², the structure functions were essentially constant.[14] The structure functions were independent of the momentum, Q^2, transferred by the incident electrons. This meant that the electrons were scattering off charged particles inside the proton. The proton contains charged particles. Experiments using deuterons, which are hydrogen-2 nuclei made of a proton and a neutron, showed that the neutron also contains charged particles.

Since the proton is made of a collection of charged particles, if it has momentum, then each particle within it carries a fraction of that total momentum. Consequently, every particle has a probability between 0 and 1 of carrying any given fraction of that momentum. If we index the particles i, and recall that the momentum fraction is x; then each particle, i, has a probability, $f_i(x)$, of carrying a fraction, x, of the proton's total momentum. Given this and letting e_i be the charge on particle i, the proton structure functions F_1 and F_2 can be expressed as functions of the momentum fraction x by

$$F_1(x) = \frac{1}{2}\sum_i e_i^2 f_i(x), \qquad (2.4)$$

and

$$F_2(x) = x\sum_i e_i^2 f_i(x). \qquad (2.5)$$

Dividing equation (2.5) into equation (2.4) gives

$$2x\frac{F_1(x)}{F_2(x)} = 1. \qquad (2.6)$$

9

The relationship between F_1 and F_2 in equation (2.6) is the Callan-Gross relation.[15] It only occurs when the particles the electron scatters off are ½ spin particles. If the proton structure functions satisfy this relation, the particles making up the proton are ½ spin particles.

The proton data from the original SLAC experiments show that for momentum fractions of $x > 0.2$, the SLAC results produce the Callan-Gross relation, indicating that the charged particles inside the proton have ½ spins.[16] This is not theory; it is an observation. The experiments show that the charged particles contained within the proton and neutron are ½ spin particles. Since the electron and the positron are charged particles with ½ spins, the SLAC experiments do not rule out the beta decay implications that the proton and the neutron are made of electrons and positrons.

3
Inside the Proton

Momentum is directly proportional to mass. If the proton is made of a collection of particles, then the fraction of its momentum carried by a particle i within it, is the ratio of the mass of the particle to the proton's mass, m_i/m_p. That is, assuming the particle does not interact with other particles within the proton. This would mean that the probability of particle i having a momentum fraction of x would be 1 at $x = m_i/m_p$, and zero everywhere else.

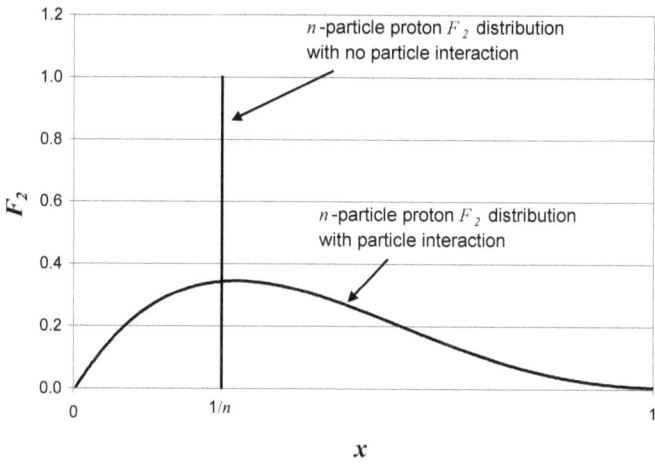

Figure 3.1: *The F_2 distributions for n particles. The distribution for interacting particles compared to that for non-interacting particles.*

If the proton were made of a collection of equally massive particles that do not interact with each other, its structure function, F_2, would only have a nonzero value where x is equal to the reciprocal of the number of particles forming the proton. For example, for a proton made of 25 equally massive particles, each particle would carry 1/25 the proton's momentum, and the structure function, F_2, would only have a nonzero value at $x = 1/25$. If there are n identical non-interacting particles making up the proton, F_2 would only have a nonzero value at $x = 1/n$, as shown in Figure 3.1.

If the particles inside the proton interact with each other - say through collisions or some kind of binding – then each particle's momentum changes over time. A particle's momentum may vary widely over time, but its average remains the average for the number of particles in the proton. This causes its momentum fraction probability to distribute around its particle-to-proton mass ratio value of x. Consequently, the F_2 curve would peak at the mass ratio value, $x = 1/n$, where n is the number of particles in the proton. However, as x goes to 0 from $1/n$, the momentum fraction probability goes to 0, causing F_2 to go to 0. As x goes to 1, the momentum fraction probability goes to 0, again causing F_2 to go to 0. Figure 3.1 also shows the F_2 curve for particles interacting within the proton.

Figure 3.2 shows a plot of the actual data collected during the original SLAC deep inelastic scattering experiments for the proton. The figure gives a fourth-order polynomial fit of the data and draws a curve of that fit through the data points. Higher-order polynomial fits do not substantially change the curve or improve the goodness of the fit. The fit curve resembles the curve in Figure 3.1 that represents interacting particles. This indicates that the particles making up the proton interact with each other.

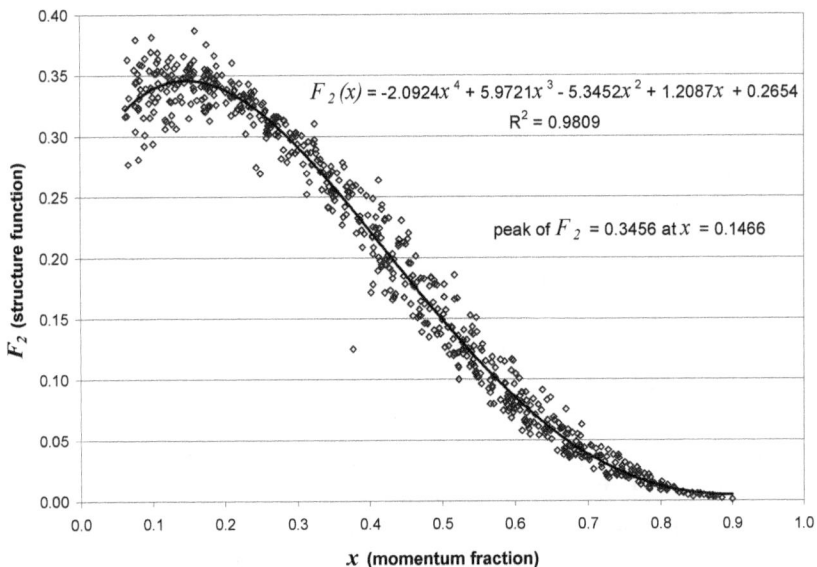

$$F_2(x) = -2.0924x^4 + 5.9721x^3 - 5.3452x^2 + 1.2087x + 0.2654$$
$$R^2 = 0.9809$$

peak of $F_2 = 0.3456$ at $x = 0.1466$

Figure 3.2: *The proton F_2 distribution measured at SLAC. The data is from experiments E49a, E49b, E61, E87, E89a and E89b.* [17]

Setting the derivative of the curve fit to zero and solving for x reveals that the curve peaks at $x = 0.1466$, with $F_2 = 0.3456$. If the momentum fraction where the F_2 peaks is the reciprocal of the number of particles forming the proton; then a peak at $x = 0.1466$ implies that the proton is made of $1/0.1466$ or about 7 interacting particles.

The proton has a mass of about 1,836 electron masses. Assuming, for simplicity's sake, that the proton is made of particles we already know exist, we are looking for a charged, ½ spin particle with a mass in the neighborhood of 262 electron masses. No known particle meets these criteria. The only charged ½ spin particles that come close to this mass are muons.[18] However, they are about 57 electron masses too light. Therefore, protons are probably not made of seven particles.

Figure 3.3 shows a plot of the actual data collected during SLAC deep inelastic scattering experiments for the deuteron. The deuteron is a hydrogen-2 nucleus made of a proton and a neutron. The data in the plot shows the per-nucleon structure function for the deuteron. In essence, the magnitude of F_2 values in this plot are half that of the values actually measured. Consequently, the curve is very similar to the proton curve shown in Figure 3.2.

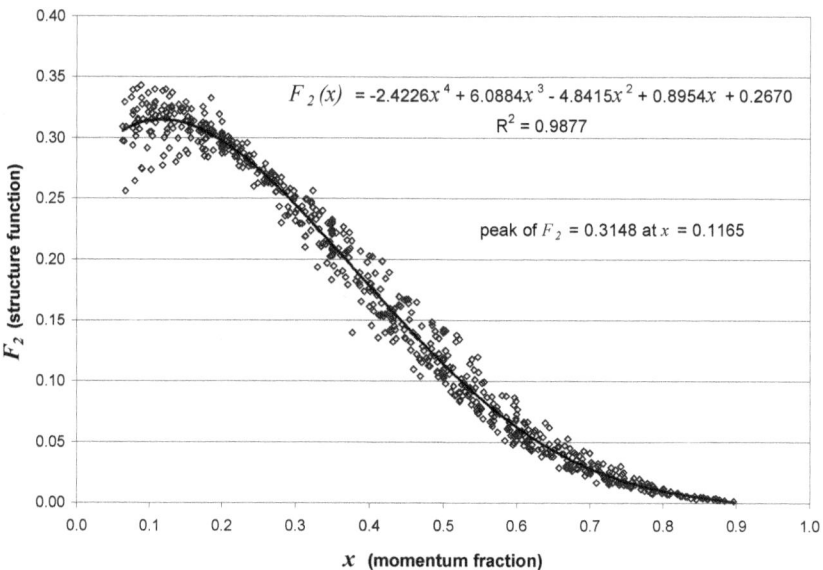

$$F_2(x) = -2.4226x^4 + 6.0884x^3 - 4.8415x^2 + 0.8954x + 0.2670$$
$$R^2 = 0.9877$$

peak of $F_2 = 0.3148$ at $x = 0.1165$

Figure 3.3: *The deuteron per-nucleon F_2 distribution measured at SLAC. The data is from experiments E49a, E49b, E61, E87, E89a, E89b, E139 and E140.* [19]

As with the proton, a fourth-order polynomial fits this data very well. When we determine the momentum fraction of the peak F_2 for this curve, we find it occurs at x = 0.1165. A peak F_2 at x = 0.1165 indicates that each nucleon in the deuteron, the proton and the neutron, is made of nine particles. For the proton to be made of nine particles, the mass of the particles must be about 204 electron masses. There is a particle with about this mass, the muon.

The muon, and its antiparticle, the antimuon, both have a free mass of about 207 electron masses. They are charged particles, the muon having a charge of -1 and the antimuon, +1; and they are both ½-spin particles. The muon appears to be a near-perfect match to the proton and neutron component particle prescribed by the deuteron deep inelastic scattering results.

The shapes of the proton F_2 curve in Figure 3.2 and the deuteron F_2 curve in Figure 3.3 indicate that the particles within the proton (and the neutron) interact with each other. The interaction is likely due to whatever is holding the particles together within the proton. For now, we assume that the same mechanism that holds the nucleons together in the nucleus holds particles together in the proton.

We know that the nucleon binding mechanism reduces the mass of the component particles it binds. Therefore, we will assume it reduces the mass of the muons inside the proton. Now, we can model the proton as a collection of nine muons held together by this binding mechanism. Nine free muons have a combined mass of 1,863 electron masses. If the binding mechanism reduces the mass of the collection of muons by about 27 electron masses, it makes the mass of our proton model about 1,836 electron masses.

The proton has a charge of +1. The charge on the muon is -1, and the charge on the antimuon is +1. In order to get a proton with a charge of +1 using muons and antimuons, our proton model must consist of one more antimuon than muon. Therefore, a proton made of a collection of nine muons and antimuons must contain four muons and five antimuons.

So, the results of the deep inelastic scattering suggests that the proton is made of four muons and five antimuons somehow bound together. There is other evidence that may support the presence of muons within protons. The electron-proton collisions in deep inelastic scattering events produce muon-antimuon pairs. Literally hundreds of the muon-antimuon pairs have been observed during these experiments.[20] Various studies characterize and explain these events within the realm of conventional knowledge.[21] Consequently, physicists attribute the reasons for these muons to other sources. However, based on this proton model, the muons and antimuons detected may be appearing because the protons shattered during these experiments are made of muons and antimuons.

Cosmic rays also appear to support the presence of muons in protons. Cosmic rays are high-energy charged particles originating from extraterrestrial sources. When they enter the Earth's atmosphere, they collide with molecules in the air and disintegrate into a variety of subatomic particles. About 87% of the charged particles in cosmic rays are protons. Most of the particles produced in the atmosphere by the cosmic rays are muons and antimuons.[22] While there are other explanations for this occurrence, it could very well be that the muons and antimuons detected come from the shattering of protons made of muons and antimuons.

4
Muon Structure

When electrons scatter off protons with momentum transfers up to 40 GeV², they reveal that the proton is made of about nine particles we assume to be muons. This is because the electron's high energy shrinks its wavelength down to a size comparable to the size of the muons making up the proton. If we continue to increase the energy of the probing electron, its wavelength will become even smaller, making it smaller than the size of the individual muons forming the proton. When this happens the electron transitions from scattering off the muons within the proton, to scattering off the particles, if any, within the muon. Now the results of the scattering produce structure functions for the muons as well as the protons.

In 1992, the Hadron Electron Ring Accelerator (HERA) began operation in Hamburg Germany. Unlike SLAC, which is a linear accelerator with fixed targets, HERA is a storage ring accelerator in which both the probe and the target move prior to the collision. In HERA, electrons with energies typically around E_e = 30 GeV, collide head on with protons of energy E_p = 820 GeV. This is the same as a 50,000 GeV electron beam hitting a fixed target in SLAC.[23] That is a wavelength of 2.5 x 10⁻²⁰ m, which is 60,000 times smaller than the diameter of the proton. At these energies, deep inelastic scattering can explore proton momentum fractions (x) over 100 times smaller than with fixed targets. It can also generate momentum transfers (Q^2) 100 times higher than those of the fixed-target experiments.

A set of experiments done at HERA in 1993 measured the proton F_2 for momentum fractions down to just under x = 2 x 10⁻⁴ and momentum transfers up to Q^2 = 1,600 GeV². They accomplished this by using an electron beam energy of E_e = 26.7 GeV and a proton energy of E_p = 820 GeV. Appendix C lists the results of those experiments.

The experiments determined F_2 values at multiple Q^2 values for each momentum fraction x. To analyze these data, we will average the values given for each momentum fraction. Just add them up and divide by the number of values. The results of this exercise are in Table 4.1. These are the results for electrons scattering off proton targets. However, because the electrons are now scattering off muons in the proton, we can produce a set of values for the muon scattering.

x	F_2
0.000178	1.187
0.000261	1.275
0.000383	1.257
0.000562	1.353
0.000825	1.320
0.001330	1.173
0.002370	1.123
0.004210	0.991
0.007500	0.815
0.013300	0.736
0.023700	0.594
0.042100	0.565
0.075000	0.508
0.133000	0.480

Table 4.1: *The results from averaging the data from the deep inelastic scattering experiments done at HERA in 1993.* [24]

First, recall that the expression for the momentum fraction of particles within the proton (equation (2.3)) is

$$x = \frac{Q^2}{2M_p \nu}. \qquad (4.1)$$

The only variable unique to the proton in the expression is the proton mass M_p. Both the Q^2 and the ν depend on the initial and final electron energies E_0 and E' (equations (2.1) and (2.2)).

Since we are now scattering off a muon within the proton, we need to get the fraction of the muon's momentum particles within it carry. We do this by simply replacing the proton mass in equation (4.1) with that of the muon. Since the muon carries one-ninth of the proton's mass, multiplying the momentum fraction for the proton by nine gives the corresponding particle momentum fraction for the muon. Table 4.2 shows the muon x values, which are the values for the proton from Table 4.1, multiplied by nine.

Next, to convert the structure function values from proton values to muon values, we need only realize that the electron begins scattering exclusively off the muon at $x = 1/9$. This is where the F_2 curve theoretically peaks for a proton made of nine particles. The momentum fraction, x, for the muon at this point is 1, so its structure function, F_2, is 0. Therefore, whatever the F_2 value is at $x = 1/9$ must be subtracted from the values at momentum fractions less than $1/9$ to get structure function values for the muon. We will use the peak value of the curve in Figure 3.2, $F_2 = 0.346$, to adjust the F_2 values in Table 4.1.

x	F_2
0.001602	0.841
0.002349	0.929
0.003447	0.911
0.005058	1.007
0.007425	0.974
0.011970	0.827
0.021330	0.777
0.037890	0.645
0.067500	0.469
0.119700	0.390
0.213300	0.248
0.378900	0.219
0.675000	0.162
1.197000	0.134

Table 4.2: *Muon structure function values converted from proton F₂ values for momentum fractions adjusted to the muon mass.* [25]

We can now use the values in Table 4.2 to plot a structure function curve for the muons inside the proton. Reviewing the data in Table 4.2, we note that the last value of x is greater than 1. This is because the original x value from Table 4.1 is greater than 0.111 (greater than $1/9$). Therefore, this point does not make the cutoff for the muon data and is not included in the plot of the muon structure function.

The third data point, $x = 0.003447$, is also excluded from the plot. The trend of the F_2 values between $x = 0.001602$ and $x = 0.005052$ is to increase. However, the F_2 value for $x = 0.003447$ is 0.911, which is less than the F_2 value of 0.929 at $x = 0.002349$. Since the decline goes against the trend of the data in that segment, the F_2 value for momentum fraction $x = 0.003447$ is not included in the plot.

Figure 4.1 shows the plot of the structure function of the muons inside the proton using the data from Table 4.2. The curve is the classic structure function curve, rising from 0 to a peak at a low momentum fraction; then declining exponentially to 0 as the momentum fraction approaches 1.

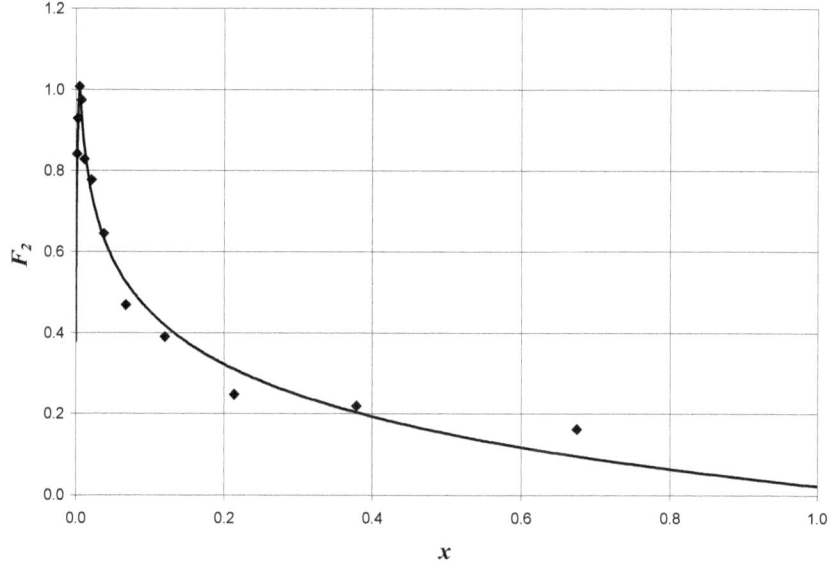

Figure 4.1: *The F_2 distribution of the muons inside the proton based on the data from the proton deep inelastic scattering experiments at HERA.* [26]

Figure 4.2 shows a blowup of the portion of the structure function curve in Figure 4.1 containing the peak. The curve is broken into two segments. The F_2 values in one segment ascend toward the peak F_2 value as the momentum fraction increases. The F_2 values in the other segment descend as the momentum fraction moves away from the peak. Both segments are well behaved and logarithmic curves track the data well. From the graph, we can see that the peak F_2 value occurs somewhere in the neighborhood of momentum fraction $x = 0.005$. The peak F_2 value is essentially 1.

We should expect a peak F_2 of 1 if the charges on the particles forming the muon are unit charges. Recall from equation (2.5) that the structure function, $F_2(x)$, can be expressed as

$$F_2(x) = x \sum_i e_i^2 f_i(x), \qquad (4.1)$$

where i denotes the individual particles, e_i, the charge on those particles, and $f_i(x)$, the probability of a particle possessing momentum fraction x. If the charges on the component particles are either $+1$ or -1, then the e_i^2 term is always 1.

The sharp peak in Figure 4.1 means that the x values of the particles cluster tightly around the x of the peak F_2. This suggests that the particles do not interact strongly. If x_a is the momentum fraction of the peak, then $1/x_a$ is the number of particles n in the muon. Because the momentum fractions carried by the particles are within a small interval $x_a \pm \delta$ of the muon's total momentum, the probability of a particle having x_a is $f_i(x_a) \sim 1$. Since $e_i^2 = 1$ and $f_i(x_a) \sim 1$ for all the particles, at momentum fraction x_a, equation (4.1) becomes

$$F_2(x_a) = x_a \sum_i 1 \times 1 = \frac{1}{n} \times n = 1. \qquad (4.2)$$

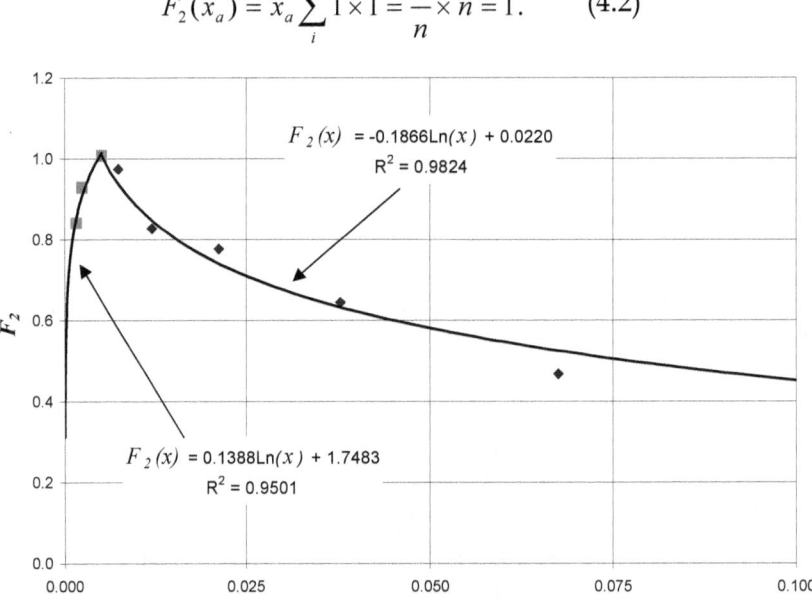

Figure 4.2: *Blowup of F_2 distribution of muons inside proton based on the data from the deep inelastic scattering experiments at HERA.* [27]

20

It appears that, ideally, the F_2 should reach a peak of 1 at the x value corresponding to the reciprocal of the number of particles forming the muon. The curve in Figure 4.2 peaks at just over 1. This near miss is likely due, in part, to the factor subtracted from the proton F_2 values in Table 4.1 to normalize them to the muon values in Table 4.2. Though it came from measured results, it is still arbitrary, in a sense. It is probably close to the correct adjustment, but there is no apparent way to know how close.

Another reason for the F_2 overshooting 1 may be that, with the data used to produce them, the two fit curves pass through $F_2 = 1$ rather than peak there. Assume both curves should reach a peak of $F_2 = 1$ at the actual momentum fraction the component particles carry. However, due to the uncertainties in the curve fits because of the small sample size of data used, they miss the mark.

One way to correct for missing the mark may be to assume that one curve's prediction of the momentum fraction that gives $F_2 = 1$ is low, and the other curve's prediction is high. Then, use as the actual $F_2 = 1$ momentum fraction, the x value midway between the two x values that make the F_2 values of the two curve fits equal 1.

The fit to the rising points in Figure 4.2 is

$$F_2(x) = 0.1388 \ln x + 1.7483, \qquad (4.3)$$

and the fit to the falling points in the figure is

$$F_2(x) = -0.1866 \ln x + 0.0220. \qquad (4.4)$$

The x value that makes $F_2 = 1$ in equation (4.3) is $x = 0.004556$. The x value that makes $F_2 = 1$ in equation (4.4) is $x = 0.005294$. The x value midway between these two points is $x = 0.004925$. This value of x now becomes our momentum fraction value for where the structure function peaks at $F_2 = 1$.

The value $x = 0.004925$ is now the momentum fraction where the peak F_2 value occurs. The reciprocal of this value is 203. This implies that the muons inside a proton are made of about 203 other particles. A muon made of 203 charged, ½ spin particles has essentially the number and type of particles needed to be a muon made of electrons.

A free muon must be made of 207 electrons to match its mass. Of course, when we say the muon is made of electrons, we really mean electrons and positrons. The muon and antimuon are charged particles. The only way to produce their charges is by using combinations of electrons possessing a unit negative charge and positrons having a unit positive charge. Since the muon has a charge of -1, and the antimuon a charge of +1, the muon must contain one more electron than positron, and the antimuon, one more positron than electron. This means that, to be made of 207 particles, the muon must contain 103 positrons and 104 electrons, and the antimuon must contain 104 positrons and 103 electrons.

The proton is made of nine muons. It has a mass of just under 1,837 electron masses. Nine muons made of 207 electrons would produce a proton with a mass of 1,863 electron masses. This is about 26 electron masses more than needed. Therefore, there appears to be a binding mechanism holding the proton together that uses about 26 electron masses or 13 MeV of energy.

The muon's structure function curve indicates that muons inside the proton are each made of about 203 electrons. If the binding mechanism holding the muons together in the proton reduces the number of electrons in the proton by 26, it leaves us with 1,837 electrons. This is just over 204 electrons per muon, very close to the number the structure function curve prescribes.

The muon model described here supports the observed decay of the muon. Muons liberated from protons decay in about 2.2 microseconds into an electron, an electron antineutrino, and a muon neutrino. The two neutrinos are virtually massless, so during the decay, all of the muon's mass but one electron converts to energy. If the free muon is the collection of 104 electrons and 103 positrons described above; then, it appears once freed from the proton, 103 pairs of electrons and positrons in it annihilate each other, leaving a lone unpaired electron as the decay product. In the antimuon, the annihilation leaves a lone positron as its decay product.

5
A Closer Look

When we look at the SLAC structure function curve for the proton (Figure 3.2), we notice that the peak F_2 value is in the neighborhood of 1/3. This may be telling us something about the internal structure of the proton.

Recall that the muon structure function curve (Figure 4.1) has a peak of approximately 1. We attribute this to the particles forming the muon all being the same size and barely interacting with each other. Consequently, we expect the F_2 value, at the x value that is the reciprocal of the number of particles in the muon, to be nearly 1. If the particles were not interacting with each other, the F_2 curve would be a single vertical line extending up to 1 at the x (see Figure 3.1). The F_2 value nearly 1 at the peak x value says that there are $1/x$ particles in the muon each carrying x fraction of its momentum.

When we have a situation like the SLAC proton F_2 curve, it is telling us that there is a mismatch involved in the curve. The F_2 peak at $x = 1/9$ tells us that the particles the electron sees have 1/9 the proton's momentum. This is because the curve is a "proton" curve. The information it presents is relative to the proton. However, the F_2 peak of 1/3 suggests that the electron is not looking at the whole proton, or else the peak would be near 1. The peak momentum fraction of 1/9 does not match the F_2 peak of 1/3.

Equation (2.3) gave the expression for calculating the momentum fraction x as

$$x = \frac{Q^2}{2M_p \nu}. \qquad (5.1)$$

This expression uses the mass of the proton, which is the target particle, in its denominator. From equations (2.1) and (2.2), we see that neither ν nor Q^2 depend on this mass. Therefore, the mass in the denominator, the target particle mass, is the only mass affecting the momentum fraction.

If we examine any of the SLAC deuteron data points listed in Appendix B, we find that, given the values for E_0, E' and Q^2 in the table, to arrive at the corresponding value of x, the proton mass must be used in equation (5.1). The momentum fraction calculations for the deuteron use the proton mass, not the mass of the target particle, the deuteron. This normalizes the SLAC deuteron F_2 curve to the proton. It is a "proton" curve. We can see this in the F_2 data for the deuteron.

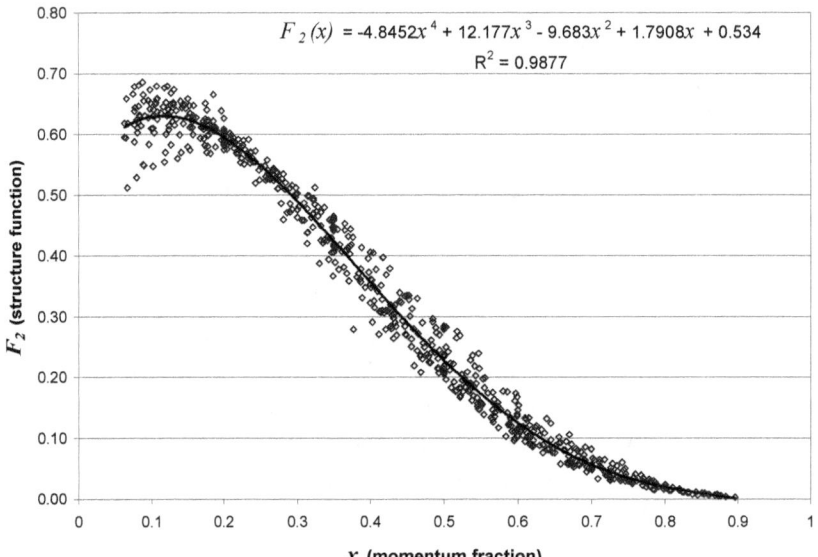

$$F_2(x) = -4.8452x^4 + 12.177x^3 - 9.683x^2 + 1.7908x + 0.534$$
$$R^2 = 0.9877$$

Figure 5.1: *The actual deuteron F_2 distribution measured at SLAC. These are the per-nucleon F_2 values multiplied by two.* [28]

Figure 3.3 shows the SLAC per-nucleon F_2 distribution for the deuteron. Per-nucleon means that the F_2 values in the graph are half of the values actually measured, since there are two nucleons in a deuteron. Figure 5.1 shows the actual F_2 distribution for the deuteron. The structure function peak is about $F_2 = 0.63$ at $x = 1/9$.[29]

A peak F_2 at $1/9$ suggests that there are only nine muons in the deuteron. However, since there are two nucleons in a deuteron, we know that there must be 18 muons in it. The reason for the discrepancy is that the curve is a proton curve, not the deuteron curve. The momentum fraction calculations use the proton's mass in them instead of the mass of the deuteron. Therefore, the calculations imply that the electron sees a particle that has $1/9$ the proton's mass.

To make this a deuteron curve, we need to calculate the momentum fractions for it using the mass of the deuteron, instead of the proton mass. We must replace the proton mass in equation 5.1 with the deuteron mass. We can do this by multiplying the proton momentum fractions by the mass ratio of the proton to the deuteron.[30]

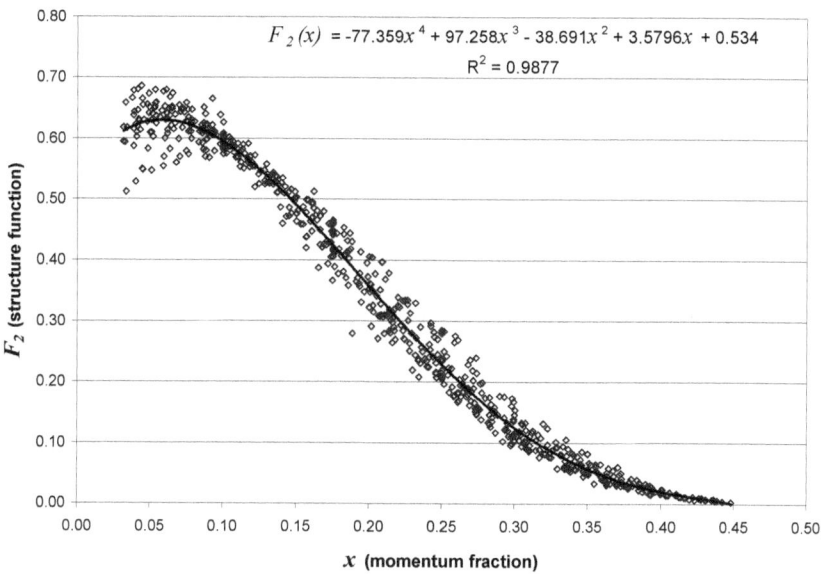

$$F_2(x) = -77.359x^4 + 97.258x^3 - 38.691x^2 + 3.5796x + 0.534$$
$$R^2 = 0.9877$$

Figure 5.2: *Deuteron F_2 curve measured at SLAC with the momentum fractions adjusted for the deuteron mass.* [31]

The mass of the proton is 0.938 GeV and the mass of the deuteron is 1.876 GeV. The ratio of these two is 0.938/1.876, or 0.5. Multiplying the momentum fractions by this ratio cuts them in half, and produces the curve shown in Figure 5.2. Now, the F_2 peaks at about a momentum fraction of 0.055 or 1/18. This indicates that the deuteron has 18 muons in it, each carrying about 1/18 of its momentum. Figure 5.2 is a "deuteron" F_2 curve.

The structure function in Figure 5.2 goes to zero at a momentum fraction of about 0.5. This is because, even though the muons are a part of the deuteron, they belong to the individual nucleons, which carry only half the deuteron's momentum. Therefore, the most momentum a muon can carry is that of its parent particle, the proton (or neutron), which is half the deuteron's momentum.

The particles the electron scatters off in proton deep inelastic scattering carry $1/9$ of the proton's momentum, which means that they are $1/9$ of the proton's mass. However, the F_2 value of about $1/3$ at $x = 1/9$ tells us that the electron is not just scattering off muons in the proton. If the electron was scattering off just individual muons in the proton, the F_2 value at $x = 1/9$ would be much closer to 1. The fact that it is not suggests that the electron is scattering off a muon that is part of a subunit of the proton larger than the muons. This means that, at the electron energies used to produce the F_2 curve in Figure 3.2, the electron is actually producing a structure function curve of those subunits, not the proton.

To understand what the curve in Figure 3.2 is telling us, recall the definition of F_2 from equation (2.5),

$$F_2(x) = x \sum_i e_i^2 f_i(x). \qquad (5.2)$$

We know that at $x = 1/9$, $F_2 \sim 1/3$; therefore,

$$\tfrac{1}{3} = \tfrac{1}{9} \sum_i e_i^2 f_i(\tfrac{1}{9}), \qquad (5.3)$$

or

$$\sum_i e_i^2 f_i(\tfrac{1}{9}) = 3. \qquad (5.4)$$

For the particles that we are dealing with, muons and antimuons, the charges are always +1 or -1. This makes e_i^2 always equal to 1. That makes equation (5.4)

$$\sum_i f_i(\tfrac{1}{9}) = 3. \qquad (5.5)$$

If we assume the probability, $f_i(x)$, is the same for all the muons and is $f(x)$, since there are nine particles in the proton, the sum in equation (5.5) becomes

$$9 \times f(\tfrac{1}{9}) = 3. \qquad (5.6)$$

The sum of the probabilities is 3. Since we are dealing with the proton, the sum is over nine particles. If the F_2 value measured were for the proton, the sum of the nine muon probabilities would be 1. The nine particles combined would carry all of the proton's momentum. Because summing over nine particles gives a value greater than 1, the F_2 value measured is not for the electron scattering off the proton. We only need sum over three particles to get a full probability of 1. Therefore, equation (5.6) suggests that the structure function curve is telling us that the electron is scattering off a unit within the proton made of only three of the nine muons.

If we sum over just three particles, the probability sum in equation (5.5) would be 1. Three particles carry all the momentum of the target off which the electron scatters. This means that during the deep inelastic scattering, the electron is now probing a subunit within the proton made of three muons, not the proton. Since there are nine muons in the proton, there are three subunits within the proton. Each of these subunits carries the full momentum that the electron sees during the deep inelastic scattering collision.

The structure function curve for the proton appears to indicate that there are three massive subunits within the proton, each carrying 1/3 of the proton's mass. Since we know there are nine muons in the proton, each subunit is apparently a cluster of three muons. Therefore, henceforth, we shall refer to the three-muon subunit of the proton as a *trimuon*. There are three trimuons in a proton: two containing two antimuons and one muon, and one made of two muons and one antimuon.

If we look again at the deuteron F_2 curve in Figure 5.2, our peak F_2 is now about 2/3, and it occurs at about $x = 1/18$. Plugging these values into equation (5.2) and remembering that e_i^2 always equals 1 for muons and antimuons, we get

$$\tfrac{2}{3} = \tfrac{1}{18} \sum_i e_i^2 f_i(\tfrac{1}{18}), \qquad (5.7)$$

or

$$\sum_i f_i(\tfrac{1}{18}) = 12. \qquad (5.8)$$

We are now summing over 18 muons since there are two nucleons in a deuteron. Note that a momentum fraction of 1/18 for the deuteron is the same momentum as a momentum fraction of 1/9 for the proton. Since the particles treated in equation (5.8) are muons, the momentum fraction probability $f_i(x)$ is the same at $x = 1/18$ for the deuteron as it is for $x = 1/9$ for the proton, $f(x)$. Therefore, the sum in equation (5.8) becomes

$$18 \times f(\tfrac{1}{18}) = 12. \qquad (5.9)$$

Given the result in equation (5.6), the expected sum in equation (5.9) is 6, not 12. Why are we getting 12? The answer seems to be that, since we have begun making this a deuteron structure function curve by adjusting the momentum fractions, we must complete the conversion by also adjusting the F_2 values.

The peak F_2 value of approximately 2/3 is for the deuteron, relative to a proton. It is the F_2 for two protons. To get the corresponding F_2 value for the deuteron, we must also apply the mass ratio of the proton to the deuteron we applied to the momentum fractions, to the F_2 values. When we do this, it cuts the F_2 values in Figure 5.2 in half. Now, we get the F_2 curve shown in Figure 5.3. The peak F_2 value in this curve is about $F_2 = 1/3$. Replacing the F_2 value in equation (5.7) with this value gives

$$\tfrac{1}{3} = \tfrac{1}{18} \sum_i e_i^2 f_i(\tfrac{1}{18}), \qquad (5.10)$$

and makes equation (5.9)

$$18 \times f(\tfrac{1}{18}) = 6. \qquad (5.11)$$

We now get a probability sum of 6 for the 18 muons contained in the two nucleons that make up the deuteron. Summing over the three muons in a trimuon gives a probability sum of 1.

The electron scattering off trimuons within the proton is similar to the electron scattering off nucleons in the deuteron. We can apply what we learned from the deuteron to convert the structure function curve of the proton shown in Figure 3.2 into a trimuon curve. We do this by multiplying both the momentum fraction values, x, and the structure function values, F_2, for the proton by the ratio of the mass of a proton to that of a trimuon. Since the trimuon is 1/3 the mass of a proton, that ratio is 3.

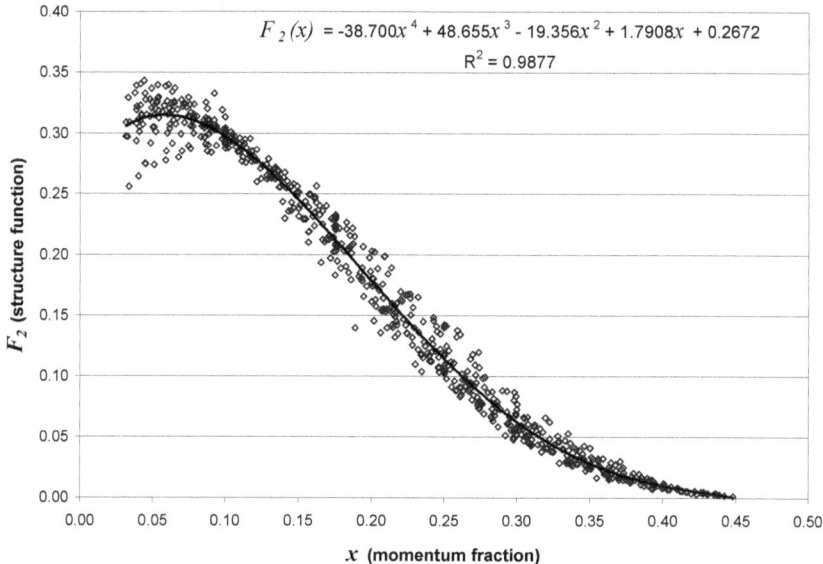

$$F_2(x) = -38.700x^4 + 48.655x^3 - 19.356x^2 + 1.7908x + 0.2672$$

$$R^2 = 0.9877$$

Figure 5.3: *Deuteron F_2 curve measured at SLAC with the momentum fractions and the F_2 values adjusted for the deuteron mass.* [32]

When we make the prescribed adjustments to the proton data and plot them, we get the resulting curve shown in Figure 5.4. Now, the trimuon structure function curve peaks at a momentum fraction of $x = 1/3$, with a value of just over 1. This is what we expect the structure function curve of the trimuon to look like. It says that the trimuon is made of three particles, muons.

The structure function peak at 1 here is not sharp as it was for the electrons in the muon (see Figure 4.1). This means that the three muons in the trimuon must interact relatively strongly. They probably bind together to some degree. Recall that our model of the muon inside the proton is about 204 electron masses, which is about 3 electron masses less than what a free muon would have. The mass deficit is probably the result of the binding mechanism that holds the muons together to form the trimuons. The binding mechanism probably causes the momentums of the muons in the trimuon to distribute over a range rather than all focus sharply at one value.

Adjusting the momentum fractions from the proton values to the trimuon values extends the possible trimuon momentum fraction values out to nearly 3. This is because the trimuon is part of the proton, which carries three times the average momentum of a trimuon. Therefore, as part of the proton system, a trimuon can possibly carry more momentum that its average.

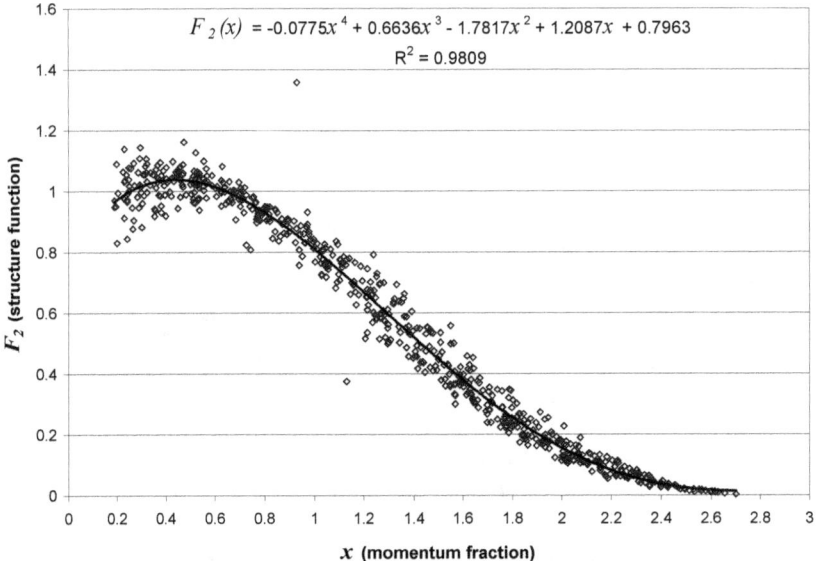

Figure 5.4: *SLAC proton F_2 curve with the momentum fractions and the F_2 values adjusted for the trimuon mass.* [33]

6
Proton Model

Our analysis of the deep inelastic scattering data from SLAC revealed that the proton is made of four muons and five antimuons. Analysis of the data from the HERA experiments showed that the muons and antimuons in the proton are made of electrons and positrons. We modeled the muon as 104 electrons and 103 positrons, and the antimuon as 103 electrons and 104 positrons. A closer look at the SLAC data indicated that the muons and antimuon within a proton cluster into subunits of three, which we call trimuons. Two of the trimuons in the proton are made of two antimuons and one muon, and the third trimuon is made of two muons and one antimuon. This is extraordinary detail about the proton.

Now that we know what pieces make up the proton, our next task, if possible, is to determine what it looks like. How do the components configure themselves within the proton? We begin building a model of the proton by modeling the muons.

We already know that the muon is a collection of about 204 electrons and positrons. Since the structure function curve in Figure 4.1 peaks sharply at 1, the electrons and positrons appear to not interact very strongly. Therefore, we shall make our muon model a spherical cluster of 104 electrons and 103 positrons, and our antimuon a cluster of 103 electrons and 104 positrons. We will assume that the particles arrange themselves in the most electrically neutral distribution.[34] This distribution balances the attractive forces on a particle from oppositely charged particles with the repulsive forces on it from like charges. This holds the muons together. Figure 6.1 shows how we depict the muon and the antimuon in this discussion.

muon   antimuon

Figure 6.1: *The depictions of the muon (left) and antimuon (right) used in the models in this discussion.*

The trimuons are particles containing three muons. Our model of the trimuon is a cluster of either two muons and one antimuon, or two antimuons and one muon. The three muons cluster in the tightest packed configuration, a triangle. Figure 6.2 shows the two forms of trimuons. To distinguish between the two, we call the configuration containing two muons a negative trimuon, and the one containing two antimuons, a positive trimuon.

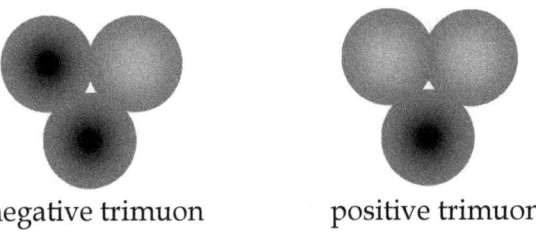

negative trimuon   positive trimuon

Figure 6.2: *The models of the negative trimuon (left) and the positive trimuon (right) used in this discussion.*

In assuming their configurations, the muons and antimuons within the trimuons make contact with each other. In our trimuon model, bonds form between two muons where they contact each other. These bonds hold the trimuon together. Three bonds form in both the negative trimuon and the positive trimuon.

In our model, the bonds between two muons in a trimuon form by the two muons sharing electrons and positrons. Recall that a free muon (and antimuon) has 207 electrons and positrons in it. However, the muons in our proton have an average of only 204 electrons and positrons. To "feel" like free muons, the muons and antimuons in the proton share particles between them to make up their deficits.

When a bond between two muons form, there is a virtual cache of particles established between the two that belongs to both muons. They each contribute to the cache; and in doing so, can claim the contents of the cache as part of their particle count. The number of particles shared between two muons depends on how many other muons with which the two are bonding.[35]

Figure 6.3 shows an example of how a bond works. In a bond between a muon and an antimuon, if the muon needs two electrons and two positrons to complete it configuration, the antimuon puts those particles into the cache. Likewise, if the antimuon needs an

electron and a positron to complete it, the muon places those particles in the cache. The two form a bond by sharing six particles.
Now, the muon has its original 203 particles plus the 4 the antimuon
put in the cache, giving it 207 particles; and the antimuon has its
original 205 particles plus the 2 the muon put into the cache, giving
it 207 particles. The catch is that because they both need the cache
to feel complete, they have to stay together. They form a bond.

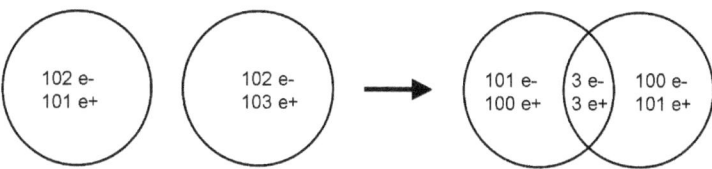

Figure 6.3: *A diagram showing how a muon and an antimuon share electrons and positrons to each feel like a free particle and form a bond between them.*

The proton is made of three trimuons, two positive trimuons and
one negative trimuon. This gives the proton 4 muons and 5 antimuons, and 919 positrons and 918 electrons. Figure 6.4 shows
diagrams of two possible configurations of the proton.

In our model of the proton, the two positive trimuons and the
negative muon bind together with three bonds. In doing so, they
form a tripod-like configuration. Each leg of the tripod has a foot
made of a muon-antimuon pair. If the tripod is wide open, the trimuons are coplanar, as in the figure on the left below; the configuration is flat and triangular. It resembles a three-bladed fan. However, the trimuons do not have to be in the same plane.

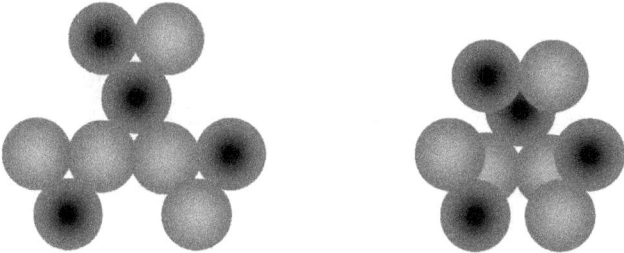

Figure 6.4: *The model of the proton containing one negative trimuon and two positive trimuons. The figure shows a planar configuration of the proton that lies flat on the page like a plate on the left and a configuration that extends out of the page like a bowl on the right.*

The model on the right in the figure above shows how the proton might look if the tripod is partially closed. It has the muon-antimuon feet rotated up off the plane of the page. This gives it more depth, and more of a 3-dimensional feel than the planar model. It resembles a table if the feet face down or a bowl if they face up. Both models have four muons and five antimuons, and both have 12 bonds.[36]

In our proton model, three bonds hold the muons in each tri-muon together. Three bonds also hold the three trimuons that form the proton together. Since nine muons are made of 1,863 electrons, but the proton only 1,837 electrons, the 12 bonds represent 26 electrons shared by the muons. If we let each muon in the trimuons share an electron-positron pair with each of its neighbors, it reduces the number of electrons in each trimuon by six. This reduces the total number of electrons in our proton model by 18. Now, if to form the proton each trimuon shares three electrons with each of its neighboring trimuons, this reduces the number of electrons in the proton by another nine. That makes the total number of electrons shared by muons in the proton, 27, one more than the 26 needed.[37]

With a mass deficit of 27 electron masses, the binding energy of the proton is about 13.5 MeV. That is about six times the binding energy of the deuteron, and just under half the binding energy of the alpha particle. This makes the proton somewhat tightly bound. The binding energy per muon is 1.5 MeV, which is greater than the binding energy per nucleon of the deuteron, but far less than the 7 MeV per nucleon with which the alpha particle binds.

Now that we have a model of the proton, let us consider the structure of its closest relative, the neutron. The neutron has a mass of 1,838.68 electron masses and a net electric charge of 0. We know that, left alone, a free neutron eventually decays into a proton and an electron. The decay also produces a neutrino and 0.78 MeV of energy. Given the complexity of the proton structure, and the out-come of the neutron decay, the neutron structure is likely only a slight modification of the proton structure.

Our proton is nine muons connected by 12 bonds. Deep inelastic scattering revealed that the deuteron is made of 18 muons. If the deuteron is a proton and a neutron, then the scattering implies that the neutron is also made of nine muons.

The mass of the proton is 1,836.16 electron masses. This is about 2.5 electron masses less than the neutron. Our model of the proton has an odd number of electrons, 1,837, in it in order to produce its positive charge. Since the neutron has no charge, its model must have an even number of electrons in it. If we round the 2.5 electron mass difference between the proton and neutron up to three whole electrons, and add them to our proton model, we get a neutron model containing 1,840 electrons, an even number.

We know that only a single electron emerges from the neutron during its decay transition to a proton. However, there is also some energy evolved. If we assume the three additional electrons that apparently make a proton into a neutron all stay together in the proton, then something can cause two of them to annihilate, producing the energy to propel the third one out of the proton. In other words, a neutron is just a proton with three additional electrons in one of its muons. To produce the scenario just described, the three electrons must be two electrons and a positron. This is consistent with the additional particles needed to neutralize the charge of a proton.

Apparently, the three electrons take up residence in one of the muons in the proton. This, of course, causes that muon to be different from a free muon. Instead of having 207 electrons like a free muon, the host muon now has 210 electrons. The muon tolerates this situation for a while, but eventually annihilates an electron-positron pair within it, producing the energy that expels the remaining extra electron from it. This brings its electron count back to 207, the number of a free muon.

Therefore, the model of our neutron is a proton with two extra electrons and an extra positron in one of its muons.[38] The internal structure of the neutron is essentially the same as that of the proton. They are the same because the neutron is a proton that has just taken on some debris that it needs to dispose of. It eventually does this through the beta decay.

7
Model Implications

In the preceding chapters, we discovered that the proton has internal structure. Not only is it not the point particle physicists thought it to be one hundred years ago; it is more than just the collection of particles previous analyses have suggested. It is a complex structure of macro-particles similar in design to the nucleus of the atom or a molecule in chemistry. That complexity provides us with a means to analyze and understand reactions involving protons (and neutrons) in a whole new light.

One type of proton reaction we can study is binding. The structure of the proton allows for at least two types of bonds to form between two of them. The first type is the single bond, shown on the left in Figure 7.1, and the second is the double bond, shown on the right in the figure. For the purposes of this discussion, we shall assume that bonds between nucleons form as a result of collisions.

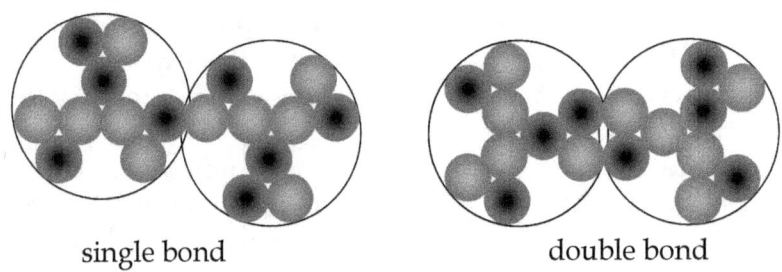

single bond   double bond

Figure 7.1: *Two protons forming a single bond between them (left), and two protons forming a double bond between them (right). In each bond, the protons share four electrons (two electron-positron pairs) between them.*

In the single bond, the collision that forms the bond causes an electron-positron pair from a muon in one proton to annihilate a pair in a muon from the other proton. This leaves both muons one electron and one positron short of having its full complement of particles. Consequently, they each share an electron-positron pair with the other to fill the void. This causes the two muons to form a bond between them similar to that shown in Figure 6.3, bonding the two protons together. The bond fills a four-electron deficit between the two protons. Therefore, the single bond creates a four-electron mass defect because of the two protons it binds together.

A similar situation exists for the double bond, except now two pairs of muons form bonds. In the collision that creates the bond, a muon-antimuon pair on the periphery of one proton collides with a pair in the other proton. The collision causes both pairs of colliding muons to lose two electron-positron pairs to annihilation. Consequently, each deficient muon borrows an electron-positron pair from its collision partner, forming a single bond between them. Since this happens for two pairs of muons, two single bonds form between the protons because of the collision. We call the matched pair of single bonds a double bond. The double bond creates an eight-electron mass defect because of the two protons bound by it, twice that of a single bond.

Using the single and double bonds, we can examine and model some of the simpler atomic nuclei.[39] The first we will consider is the deuteron. The deuteron is a hydrogen-2 nucleus, which appears to be made of a proton and a neutron. This model comes from the observation that the deuteron has the mass of two nucleons, but the charge of only one proton. Therefore, if only one of the nucleons is a proton, the other has to be a neutron.

The deuteron has a mass of 3,670.483 electron masses. This is the mass of a proton and a neutron bound together by a single bond. The proton is 1,836.153 electron masses and the neutron is 1,838.684 electron masses. Together they total 3,674.837 electron masses. If a single bond binds the two together, it reduces their mass by four electron masses. This leaves a mass of 3,670.837 electron masses, a near match to the deuteron measured mass.

We get similar results using our models of the proton and neutron. Our proton is made of 1,837 electrons and our neutron is made of 1,840 electrons. Together they total 3,677 electrons. If the two bind together with a single bond, it reduces the number of electrons in the two by four. This leaves a deuteron model containing 3,673 electrons, less than three greater than the actual number of electron masses. As with the proton and the neutron, the model mass of the deuteron is slightly higher than the actual mass.

While we do not know exactly how the two nucleons bind together to form the deuteron, the configuration likely resembles the single bond diagram in Figure 7.2. The three extra electrons in the neutron must somehow migrate to the bond site and fill in for some of the missing electrons. Otherwise, the proton would eventually expel them.

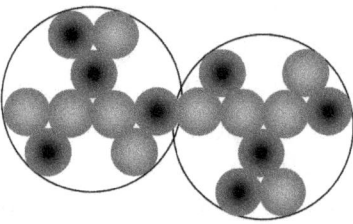

Figure 7.2: *A likely configuration of the deuteron. A single bond between two muons holds the two nucleons together.*

When the extra electrons migrate to the bond site, they replace two of the electrons and one of the positrons forming the single bond. The two bonding muons no longer need to share these particles; they can each have their own.[40] They now need to share only one positron to both feel whole. This makes the deuteron look like two protons sharing a single valence positron between them. Note that the deuteron model mass of 3,673 electron masses is just one electron mass less than the sum of two model proton masses.[41]

Next, we look at the helion, which is a helium-3 nucleus. It appears to be made of two protons and one neutron. If they pack tightly, they form a triangular configuration with three single bonds connecting them. Figure 7.3 shows two possible configurations of the helion using our proton model.

The mass of the helion is 5,495.885 electron masses. The combined mass of two protons and one neutron is 5,510.990 electron masses. Three single bonds reduce the number of electrons in the two protons and one neutron by 12, leaving a mass of 5,498.990 electron masses. This is about three electron masses greater than the measured mass.

It seems a better way to build the helion is to start with three protons. Their combined mass is 5,508.459 electron masses. If three single bonds hold them together, they reduce the mass of the protons by 12 electron masses to 5,496.459 electron masses. Now, we need only remove one positron from the group, to bring the charge down to the helium charge of +2. This leaves a helion mass of 5,495.459 electron masses, essentially the measured mass.[42]

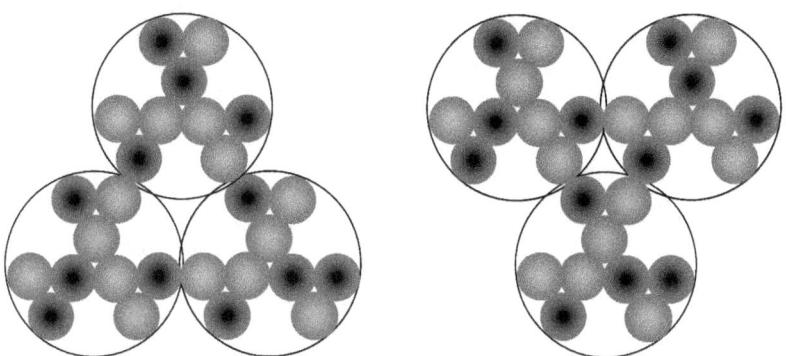

Figure 7.3: *Two possible configurations of the helion. Single bonds between two muons in each nucleon hold the three nucleons together.*

Using our model of the proton and the previous modeling scenario, combined, three of our proton models contain 5,511 electrons. Connecting them with three single bonds reduces the number of electrons to 5,499. Removing one positron to reduce the charge of the trio to +2, this leaves a helion model with 5,498 electrons. Once again, this is slightly greater than the measured mass.

Our models come close to the measured values for the masses of the particles they represent, but they are always a little high. Table 7.1 gives a tally of the discrepancies encountered so far. One reason for the differences might be that, inside the muon, the electron (and the positron) may experience a mass reduction due to whatever is holding the muon together. In other words, the electrons may suffer a mass defect inside the muon.[43]

Particle	Actual Mass	Model Mass	Difference
Proton	1,836.153	1,837	0.847
Neutron	1,838.684	1,840	1.316
Deuteron	3,670.483	3,673	2.517
Helion	5,495.885	5,498	2.115

Table 7.1: *Model masses compared to actual masses for various particles*

If we use the proton as the standard, the 1,837 electrons in our model have a mass of 1,836.153 free electron masses. This means that the mass of an electron inside the muon is only 1836.153/1,837 or 0.9995 times that of a free electron. When we apply this factor to the measured electron mass values to convert them to the number of the electrons inside the nucleons, we get much better agreement.

Of course, when converted, the proton mass is a perfect match since it is the standard used. The neutron mass goes from 1,838.684 free electrons to 1,839.604 model electrons, within one electron of our 1,840. The mass of the deuteron adjusts from 3,670.483 free electrons to 3,672.319 model electrons, and the helion from 5,495.885 free electrons to 5,498.634 model electrons. Both particle masses are now within 1 electron of our model values, 3,673 electrons and 5,498 electrons, respectively, because of the adjustment. It appears electrons inside nucleons are less massive than free electrons.

Another reason for the smaller electron mass inside the muon may be that the free electron is actually a slightly different particle. The free electron may be a complex particle composed of the electron found inside the muon and an electron neutrino. Then, the mass difference between the electron outside the muon and the electron inside the muon is the mass of the neutrino. Here, that appears to be 0.0005 electron masses or about 255 eV.

When an electron is freed from a muon via some type of decay, it may use a small portion of the energy released during the decay to create an electron neutrino-antineutrino pair. Then, the electron captures the neutrino to become a free electron, and the antineutrino flies off as a decay product.

This is likely why the electron antineutrino appears during muon decay. The electron freed during the decay must have its neutrino to be a free electron. Therefore, it produces a neutrino via pair production along with an antineutrino. The electron captures the neutrino becoming a free electron, and the antineutrino becomes a decay product.

This may also explain why, during electron capture, [44] a neutrino appears as the electron enters the nucleus. The free electron has a neutrino, but the nucleons in the nucleus will not accept it. Consequently, the electron has to shed its neutrino partner before entering the nucleus. In doing so, it carries slightly less mass into the nucleus than when it was a free electron.

Without a lot of fanfare, we will note that the triton, which is the radioactive hydrogen-3 nucleus, is just a helion with an extra electron in it. This is similar to the neutron being a proton with three extra electrons in it. The mass of the triton is 5,496.920 electron masses. It adjusts to 5,499.670 or 5,499 model electrons. This is just the 5,498-electron helion with an extra electron in it. The extra electron cancels the charge of one of the helion's valence positrons, leaving the helion with a net charge of +1, making it look like hydrogen. The helion eventually expels the electron via beta decay.[45]

The final particle we will examine is the alpha particle, which is a helium-4 nucleus. The alpha particle is made of four nucleons, seemingly two protons and two neutrons. The tightest packed arrangement of the four nucleons produces a configuration like that shown in Figure 7.4. Each nucleon contacts the other three, forming six bonds. Based on our experience with the helion, we will start our model with four protons.

The mass of an alpha particle is 7,294.300 electron masses. Adjusting this for the electron mass defect within the nucleons (dividing it by 0.9995), the alpha particle contains 7,297.948 or 7,298 electrons. Four protons have a combined 7,348 electrons. Six single bonds reduce this number by 24 electrons, down to 7,324 electrons. We must remove two positrons to bring the charge of the cluster down to +2. This leaves us with an alpha particle model containing 7,322 electrons, 24 electrons greater than the actual value.

The remedy to this problem is very simple. The bonds holding the nucleons together in the alpha particle are double bonds, not single bonds. If we replace the six single bonds with six double bonds, they reduce the number of electrons in our six-proton cluster by 48 instead of 24. This brings the number of electrons in the model down to 7,300. When we remove the two positrons to adjust the charge, our model now has 7,298 electrons.

Figure 7.4: *The likely arrangement of the nucleons forming the alpha particle.*

Looking back at Figure 7.1, we see that in order to form the alpha particle modeled above, each nucleon uses all three of its double bonds to bond with its three companion nucleons. Consequently, the protons that go into forming the alpha particle suffer a significant loss of electrons to form it. This causes the alpha particle to have a large mass defect and a high binding energy per nucleon.

There is one question we might ask after looking at how the three preceding particles are modeled. Since single bonds hold the deuteron and the triton (helion) together, but double bonds hold the alpha particle together, would one expect deuterium or tritium to fuse easily into helium-4? Seeing the radical transformation that must take place within the protons to go from a few single bonds to six double bonds, the answer seems to be – no. Our experience with controlled thermonuclear fusion seems to support this answer.

The internal structure of the proton makes putting four of them together to form an alpha particle much more complicated than gluing four spheres together with some magical force. [46] Furthermore, it seems trying to do it on a small scale at high energies would be even more difficult. Getting the muons to align properly prior to the collisions that form the bonds does not seem easy to do.[47]

We can now make an interesting observation about the make up of the universe. In our model, the nucleons are made of muons and antimuons, and muons and antimuons are made of electrons and positrons. Protons have one more positron than electron in them and neutrons have the same number of electrons and positrons in them. Therefore, if we include the orbital electrons in an atom, neutral atoms contain equal numbers of electrons and positrons.

Since electrons are matter particles and positrons are their antimatter particles, our proton model implies that neutral atoms contain equal amounts of matter and antimatter.[48] If the universe is electrically neutral, then it contains as much antimatter as it does matter. Everything in the universe is a perfect balance of matter and antimatter, as it should be.

8
Summary

In the preceding chapters, we took an amazing journey through the inside of the proton. We began by noticing that radioactive beta decay seems to indicate that the nucleus of the atom contains electrons and positrons. We saw that, over time, some atomic nuclei emit several negative or positive beta particles to become stable. These beta particles have the same charges and masses as electrons and positrons. Consequently, electrons and positrons appear to exit the nucleus.

Next, we explored the realm of deep inelastic scattering, where very high-energy electrons are fired into protons targets. The scattering energies and angles of the electrons produce a picture of what the inside of the proton looks like. The first deep inelastic scattering experiments were done at the Stanford Linear Accelerator Center (SLAC) in 1967. They revealed that the electrons scattered off charged particles inside the proton, and that those particles were ½ spin particles.

Then we analyzed the structure function data for the proton collected by SLAC. The structure function is an indicator of how likely a particle inside the proton is of carrying a certain amount of the proton's momentum. It is a function of the momentum fraction of the particles within the proton. The structure function peaks at the momentum fraction the particles within the proton carries. The reciprocal of the peak momentum fraction is the number of particles the proton contains.

The SLAC structure function curves for the proton peaked at about 0.11 (1/9) indicating that the proton is made of nine particles. Based on the mass of the proton, we concluded that the particles inside the proton are muons. We determined that to produce its mass and charge, the proton must be made of four muons and five antimuons.

The proton structure function was examined further using data generated in 1993 by the Hadron Electron Ring Accelerator (HERA).

The HERA scatterings occurred at energies significantly higher than the energies of the SLAC experiments. They provided a much higher resolution of the inside of the proton and showed that the muons making up the proton were also made of component particles. The HERA data looked at proton momentum fractions less than the peak of 1/9 the SLAC data revealed. When adjusted to the mass of the muon, the peak of the HERA structure function curve indicates that the muon is made of about 203 charged ½ spin particles. Since the muon's mass is about 207 times that of an electron, we concluded that the particles within the muons and antimuons are apparently electrons and positrons.

We noticed that the structure function curve for the electrons in the muon peaked at 1, but the structure function curve for the muons in the proton only rose to about 1/3. This led us to realize that, while the electrons were scattering off muons in the proton, those muons were part of a larger substructure within the proton.

Using the definition of the structure function, we determined that the muons in the proton cluster into groups of three that we call trimuons. The proton is made of three trimuons. There are two types of trimuons. One type is made of two muons and one antimuon. We call it a negative trimuon. The other is made of two antimuons and one muon. We call it a positive trimuon.

To produce the model of the proton, we started by modeling the muons and antimuons. The electrons and positrons that make up the muons and antimuons must coexist within them. For this to occur, we assumed they arrange themselves in a way that balances the attractive and repulsive electrical forces between them. This allows them to reside within the muons without annihilating each other.

We recognized that to have trimuons, a mechanism must exist that binds muons together. We recalled that, while a free muon must be made of 207 particles, the muon structure function curve indicated that the muons inside the proton are made of only about 203 particles. We concluded that the muons inside the proton contain about 204 particles on average, but they all "feel" as if they contain 207 particles because they share particles between themselves. Each muon shares three or four particles with its neighbor to help it reach 207 particles. Because they share particles, the two muons must stay together. This creates a bond between them. Three bonds hold the three muons together.

44

Summary

We used the same mechanism that binds the muons together into trimuons to bind trimuons together into the proton. One muon from each trimuon is bound to a muon from the other two trimuons, forming a tripod-like arrangement. We determined that the neutron is just a proton with two extra electrons and one extra positron in one of its muons. The muon eventually expels the extra particles, showing the nucleon as the proton that it is.

Based on its configuration, we see that protons can form at least two types of bonds with other nucleons. We call the first type a single bond. A single bond is similar to the bonds that bind muons together to form the proton. In it, a muon from each nucleon shares four particles. Sharing the particles causes the two nucleons to have to stay together. The single bond creates a mass defect of four electron masses.

The second type of bond is the double bond. As it implies, in it two muons from one proton bind with two muons from the other proton. Again, each muon pair shares four particles, so the two protons share eight particles in all. The double bond creates a mass defect of eight electron masses.

With these bonds, we were able to show that the deuteron is made of a proton and a neutron connected by a single bond. We concluded that the extra particles in the neutron must migrate to the site of the single bond and relieve three of the particle there to avoid being eventually expelled. This leaves the deuteron looking like two protons sharing a single valence positron. We were also able to show that the helion is made of three protons bound together by three single bonds, less one valence positron to adjust the charge. The triton turned out to be a helion with an extra electron in one of its muons. The helion eventually expels the electron via beta decay.

At this point, we noticed that the number of electrons in our models were always slightly more than the actual electron-mass values of the particles modeled. We hypothesized that the electrons and positrons that made up the muons inside the proton suffer a mass defect as a result of being part of the muon. We used the proton as our standard and generated a mass-defect factor of 0.9995 for electrons inside a muon. This gave a proton containing 1,837 electrons a mass of 1,836.153 free electron masses. When applied to the other particles, their electron mass values all are within one of the number of electrons in our models.

We further hypothesized that a free electron may be a complex particle made of the electron found inside the muon and an electron neutrino. The neutrino is the difference in mass between an electron inside the muon and one outside the muon. Inside the muon, the electron has no neutrino, but outside the muon, it must have one. During decay, the outgoing electron uses some of the decay energy to pair produce a neutrino and antineutrino. The exiting electron captures the neutrino to become a free electron, and the antineutrino flies off as a decay product.

We then showed that, in order to produce the alpha particle model, six double bonds must bind the four nucleons forming it. The four particles cluster into their tightest packed formation and each uses all three of its potential double bonds to bond with its three neighbors. The six bonds create a mass defect of 48 electron masses, and give the alpha particle a very high binding energy per nucleon.

Upon seeing the difference between the bond structures of the deuteron and the helion/triton versus that of the alpha particle, we questioned whether we should expect deuterons and tritons to fuse easily into alpha particles. It probably is not easy to get the single-bonded deuterons and tritons to reconfigure into double-bonded particles, especially at the high energies used in controlled thermonuclear fusion.

Finally, we recognized that, if muons and antimuons are made of electrons and positrons; and protons and neutrons are made of muons and antimuons; then all matter is made of electrons and positrons. Furthermore, since there is an unpaired positron in every proton; and a matching electron in the electron cloud of every neutral atom; with neutrons having an equal number of electrons and positrons; there appears to be an equal amount of matter and antimatter in the universe. Nature just camouflages it inside the nucleons of the atom.

The recognition that the nucleus is made of electron and positrons takes us back to our original observation of the radioactive isotopes decaying via beta emission. What appeared to be electrons and positrons emanating from the nucleus, actually turned out to be electrons and positrons emerging from the nucleus. Sometimes things are what they appear to be.

Appendices

Appendix A
SLAC Proton DIS Data

This appendix contains the final cross sections and the F_2 values extracted from them from the analysis of the SLAC deep inelastic scattering data for the proton. The data is from Files E.2 and E.4 of the Ph.D. thesis for Stanford University, written by L.W. Whitlow entitled:

Deep Inelastic Structure Functions
from Electron Scattering on Hydrogen,
Deuterium, and Iron at 0.6 GeV² < Q^2 < 30.0 GeV²,

document number SLAC-357, dated March 1990.

In the table:

I	=	counting index
J	=	code for SLAC experiment number:
		1 = E49a, 2 = E49b, 3 = E61,
		4 = E87, 5 = E89a, 6 = E89b
E_0	=	incident beam energy (GeV)
E'	=	scattered electron energy (GeV)
θ	=	scattering angle (degrees)
x	=	Bjorken scaling variable (momentum fraction)
Q^2	=	4-momentum transfer squared (GeV²)
σ	=	final measured cross section (pb/srGeV)
F_2	=	final extracted structure function

Details concerning the collection, analysis, preparation and error treatment of this data are in the reference given above.

Table A.1: SLAC Proton DIS Data

I	J	E_0	E'	θ	x	Q^2	σ	F_2
1	1	10.027	8.27	5.988	0.274	0.904	1.20E+06	0.30016
2	1	10.027	7.955	5.988	0.224	0.87	1.10E+06	0.32066
3	1	10.027	7.542	5.988	0.177	0.824	9.32E+05	0.32355
4	1	10.027	7.257	5.988	0.153	0.794	8.52E+05	0.32693
5	1	10.027	6.731	5.988	0.119	0.737	6.98E+05	0.3122
6	1	10.027	6.088	5.988	0.09	0.666	5.82E+05	0.30119
7	1	10.027	5.351	5.988	0.067	0.586	4.75E+05	0.27682
8	1	13.549	11.396	5.988	0.417	1.685	4.28E+05	0.23967
9	1	13.549	11.146	5.988	0.365	1.647	4.41E+05	0.2749
10	1	13.549	10.837	5.988	0.315	1.601	4.22E+05	0.29547
11	1	13.549	10.502	5.988	0.271	1.552	4.02E+05	0.31447
12	1	13.549	9.844	5.988	0.209	1.455	3.54E+05	0.33159
13	1	13.549	9.218	5.988	0.168	1.363	3.21E+05	0.34548
14	1	13.549	8.492	5.988	0.132	1.256	2.69E+05	0.32928
15	1	13.549	7.715	5.988	0.104	1.141	2.40E+05	0.32845
16	1	13.549	6.889	5.988	0.081	1.019	2.18E+05	0.32385
17	1	16.075	13.554	5.988	0.503	2.377	1.93E+05	0.17809
18	1	16.075	13.309	5.988	0.45	2.334	2.09E+05	0.21125
19	1	16.075	13.009	5.988	0.396	2.281	2.25E+05	0.25099
20	1	16.075	12.668	5.988	0.347	2.221	2.23E+05	0.27517
21	1	16.075	12.072	5.988	0.282	2.116	2.18E+05	0.31261
22	1	16.075	11.373	5.988	0.226	1.995	2.01E+05	0.33441
23	1	16.075	10.67	5.988	0.185	1.872	1.78E+05	0.33393
24	1	16.075	9.908	5.988	0.15	1.738	1.67E+05	0.34879
25	1	16.075	9.087	5.988	0.122	1.594	1.50E+05	0.34326
26	1	16.075	8.208	5.988	0.098	1.44	1.38E+05	0.34013
27	1	16.075	7.27	5.988	0.077	1.275	1.32E+05	0.34363
28	1	16.075	6.478	5.988	0.063	1.136	1.21E+05	0.32321
29	1	19.544	16.414	5.988	0.596	3.5	6.87E+04	0.11615
30	1	19.544	16.158	5.988	0.542	3.446	7.84E+04	0.14308
31	1	19.544	15.864	5.988	0.49	3.383	9.03E+04	0.17846
32	1	19.544	15.54	5.988	0.441	3.313	1.00E+05	0.21499
33	1	19.544	14.967	5.988	0.372	3.191	1.04E+05	0.25352
34	1	19.544	14.287	5.988	0.309	3.047	1.06E+05	0.29241
35	1	19.544	13.25	5.988	0.239	2.818	1.00E+05	0.32459
36	1	19.544	12.046	5.988	0.183	2.569	9.27E+04	0.34746
37	1	19.544	11.171	5.988	0.152	2.382	8.82E+04	0.35876
38	1	19.544	10.261	5.988	0.126	2.188	8.37E+04	0.36489
39	1	19.544	9.455	5.988	0.107	2.017	7.76E+04	0.35363
40	1	19.544	8.65	5.988	0.09	1.845	7.73E+04	0.36392
41	1	19.544	7.774	5.988	0.075	1.658	7.39E+04	0.35448
42	1	7.019	5.18	10	0.32	1.104	2.86E+05	0.27664
43	1	7.019	4.941	10	0.27	1.053	2.79E+05	0.30144
44	1	7.019	4.762	10	0.24	1.016	2.66E+05	0.30967
45	1	7.019	4.276	10	0.177	0.912	2.38E+05	0.32512
46	1	7.019	3.71	10	0.127	0.791	2.04E+05	0.31725
47	1	7.019	3.082	10	0.089	0.657	1.73E+05	0.29204
48	1	9.022	6.824	10	0.454	1.871	1.15E+05	0.22037
49	1	9.022	6.353	10	0.348	1.741	1.18E+05	0.27128
50	1	9.022	5.358	10	0.214	1.469	1.13E+05	0.33795
51	1	9.022	4.069	10	0.12	1.115	9.45E+04	0.337
52	1	10.998	8.268	10	0.539	2.762	4.32E+04	0.1526
53	1	10.998	8.062	10	0.489	2.693	4.81E+04	0.18149

Table A.1: SLAC Proton DIS Data

I	J	E_0	E'	θ	x	Q^2	σ	F_2
54	1	10.998	7.83	10	0.44	2.616	5.23E+04	0.21158
55	1	10.998	7.467	10	0.376	2.494	5.67E+04	0.25243
56	1	10.998	6.875	10	0.297	2.297	5.97E+04	0.30233
57	1	10.998	6.279	10	0.237	2.098	5.96E+04	0.33396
58	1	10.998	5.627	10	0.187	1.88	5.62E+04	0.3422
59	1	10.998	4.919	10	0.144	1.644	5.33E+04	0.34421
60	1	10.998	4.155	10	0.108	1.388	5.02E+04	0.33206
61	1	10.998	3.335	10	0.077	1.114	5.06E+04	0.32474
62	1	13.545	10.081	10	0.638	4.148	1.34E+04	0.091
63	1	13.545	9.829	10	0.58	4.045	1.61E+04	0.11591
64	1	13.545	9.696	10	0.553	3.991	1.68E+04	0.12534
65	1	13.545	9.243	10	0.471	3.804	2.23E+04	0.18311
66	1	13.545	8.733	10	0.398	3.594	2.61E+04	0.23489
67	1	13.545	8.157	10	0.332	3.357	2.85E+04	0.28093
68	1	13.545	7.514	10	0.273	3.092	2.89E+04	0.30827
69	1	13.545	6.849	10	0.224	2.819	2.91E+04	0.33113
70	1	13.545	6.096	10	0.179	2.509	2.85E+04	0.34092
71	1	13.545	5.32	10	0.142	2.189	2.84E+04	0.34797
72	1	13.545	4.456	10	0.108	1.834	2.99E+04	0.36327
73	1	13.545	3.746	10	0.084	1.542	2.91E+04	0.33888
74	1	15.204	11.213	10	0.692	5.18	6.40E+03	0.06271
75	1	15.204	10.787	10	0.601	4.982	9.73E+03	0.10437
76	1	15.204	10.383	10	0.53	4.796	1.21E+04	0.14042
77	1	15.204	9.856	10	0.454	4.553	1.55E+04	0.19577
78	1	15.204	9.296	10	0.387	4.294	1.69E+04	0.2311
79	1	15.204	8.688	10	0.328	4.014	1.92E+04	0.28058
80	1	15.204	8.007	10	0.274	3.699	1.95E+04	0.30358
81	1	15.204	7.301	10	0.227	3.373	2.04E+04	0.33368
82	1	15.204	6.522	10	0.185	3.013	2.16E+04	0.36625
83	1	15.204	5.695	10	0.147	2.631	2.05E+04	0.35421
84	1	15.204	4.989	10	0.12	2.305	2.09E+04	0.35712
85	1	15.204	4.259	10	0.096	1.968	2.15E+04	0.35394
86	1	17.706	12.777	10	0.743	6.873	2.16E+03	0.03517
87	1	17.706	12.509	10	0.69	6.729	3.12E+03	0.05328
88	1	17.706	12.356	10	0.662	6.647	3.38E+03	0.05917
89	1	17.706	11.972	10	0.599	6.441	5.16E+03	0.09594
90	1	17.706	11.477	10	0.528	6.174	6.54E+03	0.13006
91	1	17.706	10.956	10	0.465	5.894	8.50E+03	0.17991
92	1	17.706	10.352	10	0.404	5.569	9.66E+03	0.21761
93	1	17.706	9.693	10	0.347	5.215	1.07E+04	0.25488
94	1	17.706	9.006	10	0.297	4.845	1.15E+04	0.28623
95	1	17.706	8.265	10	0.251	4.446	1.21E+04	0.31369
96	1	17.706	7.441	10	0.208	4.003	1.25E+04	0.33321
97	1	17.706	6.782	10	0.178	3.649	1.29E+04	0.34573
98	1	17.706	6.069	10	0.15	3.265	1.33E+04	0.35661
99	1	17.706	5.3	10	0.122	2.851	1.38E+04	0.36177
100	1	17.706	4.531	10	0.099	2.438	1.53E+04	0.38122
101	1	17.706	3.735	10	0.077	2.009	1.66E+04	0.37937
102	1	19.35	13.722	10	0.764	8.067	1.12E+03	0.02465
103	1	19.35	13.462	10	0.716	7.914	1.61E+03	0.03698
104	1	19.35	13.023	10	0.645	7.656	2.71E+03	0.06626
105	1	19.35	12.498	10	0.571	7.348	3.95E+03	0.10294
106	1	19.35	11.969	10	0.508	7.037	5.17E+03	0.14264
107	1	19.35	11.381	10	0.447	6.691	6.33E+03	0.18475

Table A.1: SLAC Proton DIS Data

I	J	E_0	E'	θ	x	Q^2	σ	F_2
108	1	19.35	10.763	10	0.393	6.328	7.15E+03	0.21957
109	1	19.35	10.087	10	0.341	5.931	8.16E+03	0.26182
110	1	19.35	9.352	10	0.293	5.498	8.62E+03	0.28716
111	1	19.35	8.558	10	0.248	5.032	9.25E+03	0.31681
112	1	19.35	7.882	10	0.215	4.634	9.20E+03	0.31944
113	1	19.35	7.176	10	0.185	4.219	9.95E+03	0.34692
114	1	19.35	6.47	10	0.157	3.804	1.02E+04	0.35328
115	1	19.35	5.706	10	0.131	3.355	1.01E+04	0.33947
116	1	19.35	4.912	10	0.107	2.888	1.14E+04	0.36865
117	1	19.35	4.059	10	0.083	2.386	1.21E+04	0.359
118	2	4.504	2.653	18.02	0.337	1.171	7.14E+04	0.27603
119	2	4.504	2.499	18.02	0.293	1.103	7.30E+04	0.29838
120	2	4.504	2.363	18.02	0.26	1.043	7.33E+04	0.31265
121	2	4.504	2.25	18.02	0.235	0.994	7.21E+04	0.31662
122	2	4.504	2	18.02	0.188	0.884	7.09E+04	0.32581
123	2	4.504	1.75	18.02	0.15	0.773	6.73E+04	0.31477
124	2	4.504	1.5	18.02	0.118	0.663	6.64E+04	0.3061
125	2	6.509	3.939	18.02	0.521	2.514	1.48E+04	0.16653
126	2	6.509	3.826	18.02	0.485	2.443	1.59E+04	0.18411
127	2	6.509	3.556	18.02	0.409	2.269	1.88E+04	0.23344
128	2	6.509	3.325	18.02	0.355	2.121	2.08E+04	0.26983
129	2	6.509	3.05	18.02	0.3	1.946	2.17E+04	0.2936
130	2	6.509	2.782	18.02	0.253	1.772	2.36E+04	0.32733
131	2	6.509	2.424	18.02	0.202	1.545	2.33E+04	0.32751
132	2	6.509	2.242	18.02	0.179	1.43	2.32E+04	0.3251
133	2	6.509	2.061	18.02	0.157	1.313	2.44E+04	0.33791
134	2	6.509	1.5	18.02	0.102	0.958	2.59E+04	0.32012
135	2	8.614	5.087	18.02	0.649	4.297	3.20E+03	0.08494
136	2	8.614	4.942	18.02	0.606	4.175	3.69E+03	0.10065
137	2	8.614	4.698	18.02	0.54	3.967	4.86E+03	0.13808
138	2	8.614	4.422	18.02	0.475	3.733	6.18E+03	0.18281
139	2	8.614	4.105	18.02	0.409	3.465	7.33E+03	0.2246
140	2	8.614	3.749	18.02	0.346	3.163	8.54E+03	0.26887
141	2	8.614	3.435	18.02	0.298	2.899	9.50E+03	0.30317
142	2	8.614	3.209	18.02	0.267	2.712	9.71E+03	0.31102
143	2	8.614	2.96	18.02	0.236	2.499	1.05E+04	0.3363
144	2	8.614	2.735	18.02	0.209	2.308	1.09E+04	0.34507
145	2	8.614	2.545	18.02	0.189	2.149	1.13E+04	0.35001
146	2	8.614	2.327	18.02	0.167	1.966	1.12E+04	0.34082
147	2	8.614	2.157	18.02	0.15	1.823	1.20E+04	0.35465
148	2	8.614	2	18.02	0.136	1.69	1.20E+04	0.34317
149	2	8.614	1.501	18.02	0.095	1.268	1.33E+04	0.32703
150	2	10.392	5.998	18.02	0.742	6.115	7.91E+02	0.0375
151	2	10.392	5.75	18.02	0.673	5.862	1.20E+03	0.05873
152	2	10.392	5.498	18.02	0.61	5.605	1.76E+03	0.08898
153	2	10.392	5.25	18.02	0.555	5.352	2.37E+03	0.12325
154	2	10.392	4.767	18.02	0.46	4.857	3.43E+03	0.18538
155	2	10.392	4.44	18.02	0.405	4.522	4.10E+03	0.2251
156	2	10.392	4.026	18.02	0.343	4.098	4.85E+03	0.26848
157	2	10.392	3.7	18.02	0.3	3.768	5.22E+03	0.28827
158	2	10.392	3.423	18.02	0.266	3.485	5.66E+03	0.30956
159	2	10.392	3.192	18.02	0.241	3.254	6.24E+03	0.33705
160	2	10.392	2.912	18.02	0.211	2.962	6.50E+03	0.34308
161	2	10.392	2.688	18.02	0.19	2.74	6.67E+03	0.34348

Table A.1: SLAC Proton DIS Data

I	J	E_0	E'	θ	x	Q^2	σ	F_2
162	2	10.392	2.497	18.02	0.172	2.546	6.63E+03	0.332
163	2	10.392	2.317	18.02	0.156	2.362	7.22E+03	0.35096
164	2	10.392	2.136	18.02	0.141	2.178	7.02E+03	0.32804
165	2	10.392	1.997	18.02	0.129	2.036	7.64E+03	0.34459
166	2	10.392	1.792	18.02	0.113	1.827	8.33E+03	0.35348
167	2	10.392	1.497	18.02	0.091	1.526	8.98E+03	0.33938
168	2	12.518	5.914	18.02	0.586	7.263	1.04E+03	0.09671
169	2	12.518	4.947	18.02	0.428	6.075	2.07E+03	0.19787
170	2	12.518	4.121	18.02	0.321	5.061	2.94E+03	0.2768
171	2	12.518	3.314	18.02	0.236	4.07	3.74E+03	0.33135
172	2	12.518	2.944	18.02	0.201	3.615	3.97E+03	0.33628
173	2	12.518	2.689	18.02	0.179	3.302	4.33E+03	0.3529
174	2	12.518	2.428	18.02	0.157	2.982	4.67E+03	0.36228
175	2	12.518	2.249	18.02	0.143	2.762	4.90E+03	0.36571
176	2	12.518	2.07	18.02	0.13	2.542	4.99E+03	0.35588
177	2	12.518	1.886	18.02	0.116	2.316	5.23E+03	0.35139
178	2	12.518	1.746	18.02	0.106	2.144	5.64E+03	0.36051
179	2	12.518	1.605	18.02	0.096	1.971	5.85E+03	0.35402
180	2	12.518	1.365	18.02	0.08	1.669	6.57E+03	0.35137
181	2	13.32	7	18.02	0.771	9.147	2.04E+02	0.02169
182	2	13.32	6.5	18.02	0.664	8.494	5.40E+02	0.05966
183	2	13.32	5.838	18.02	0.543	7.619	1.02E+03	0.11629
184	2	13.32	5.382	18.02	0.471	7.021	1.43E+03	0.16418
185	2	13.32	4.894	18.02	0.404	6.382	1.89E+03	0.21679
186	2	13.32	4.338	18.02	0.336	5.66	2.40E+03	0.2703
187	2	13.32	3.928	18.02	0.291	5.125	2.71E+03	0.29717
188	2	13.32	3.603	18.02	0.258	4.699	2.87E+03	0.30564
189	2	13.32	3.344	18.02	0.233	4.37	3.31E+03	0.34292
190	2	13.32	3.008	18.02	0.203	3.919	3.41E+03	0.33864
191	2	13.32	2.689	18.02	0.176	3.509	3.60E+03	0.33939
192	2	13.32	2.5	18.02	0.161	3.267	3.61E+03	0.32951
193	2	13.32	2.312	18.02	0.146	3.021	3.99E+03	0.34852
194	2	13.32	1.972	18.02	0.121	2.573	4.39E+03	0.34901
195	2	13.32	1.786	18.02	0.108	2.334	4.61E+03	0.34363
196	2	13.32	1.65	18.02	0.098	2.153	5.23E+03	0.36937
197	2	13.32	1.5	18.02	0.088	1.96	4.77E+03	0.31517
198	2	17.027	8	18.02	0.789	13.363	6.43E+01	0.01494
199	2	17.027	7.5	18.02	0.701	12.528	1.48E+02	0.0349
200	2	17.027	7	18.02	0.621	11.693	2.73E+02	0.06476
201	2	17.027	6.5	18.02	0.55	10.858	4.27E+02	0.1015
202	2	17.027	5.885	18.02	0.469	9.816	6.73E+02	0.15756
203	2	17.027	5.432	18.02	0.417	9.069	8.38E+02	0.19276
204	2	17.027	4.929	18.02	0.362	8.22	1.02E+03	0.22701
205	2	17.027	4.383	18.02	0.308	7.302	1.30E+03	0.27709
206	2	17.027	4	18.02	0.273	6.682	1.48E+03	0.30301
207	2	17.027	3.75	18.02	0.251	6.264	1.64E+03	0.32609
208	2	17.027	3.42	18.02	0.223	5.701	1.71E+03	0.32333
209	2	17.027	3.084	18.02	0.196	5.139	1.96E+03	0.35041
210	2	17.027	2.75	18.02	0.171	4.594	1.93E+03	0.32435
211	2	17.027	2.5	18.02	0.153	4.176	2.18E+03	0.34563
212	2	17.027	2.226	18.02	0.134	3.718	2.43E+03	0.35916
213	2	17.027	2	18.02	0.118	3.341	2.27E+03	0.31387
214	2	17.027	1.8	18.02	0.105	3.002	2.73E+03	0.35163
215	2	17.027	1.5	18.02	0.086	2.506	3.19E+03	0.35834

Table A.1: SLAC Proton DIS Data

I	J	E_0	E'	θ	x	Q^2	σ	F_2
216	2	4.506	2.2	26.015	0.464	2.009	1.09E+04	0.20143
217	2	4.506	2	26.015	0.388	1.826	1.33E+04	0.25516
218	2	4.506	1.8	26.015	0.324	1.644	1.52E+04	0.29701
219	2	4.506	1.6	26.015	0.268	1.461	1.61E+04	0.31615
220	2	4.506	1.4	26.015	0.219	1.278	1.67E+04	0.32192
221	2	4.506	1.2	26.015	0.177	1.096	1.91E+04	0.35103
222	2	4.506	1	26.015	0.139	0.913	1.96E+04	0.33279
223	2	6.711	3	26.015	0.586	4.08	1.85E+03	0.11441
224	2	6.711	2.75	26.015	0.503	3.74	2.67E+03	0.1676
225	2	6.711	2.442	26.015	0.414	3.315	3.57E+03	0.22426
226	2	6.711	2.25	26.015	0.366	3.06	3.88E+03	0.2419
227	2	6.711	2.055	26.015	0.319	2.788	4.67E+03	0.28654
228	2	6.711	1.89	26.015	0.284	2.57	5.15E+03	0.30997
229	2	6.711	1.75	26.015	0.256	2.38	5.48E+03	0.32147
230	2	6.711	1.631	26.015	0.233	2.218	5.66E+03	0.32457
231	2	6.711	1.5	26.015	0.209	2.04	5.72E+03	0.31706
232	2	6.711	1.381	26.015	0.188	1.878	6.73E+03	0.358
233	2	6.711	1.19	26.015	0.156	1.618	6.91E+03	0.34024
234	2	6.711	1	26.015	0.127	1.36	7.02E+03	0.31027
235	2	8.713	3.751	26.015	0.711	6.623	3.18E+02	0.04272
236	2	8.713	3.502	26.015	0.632	6.183	5.85E+02	0.0792
237	2	8.713	3.252	26.015	0.56	5.742	8.50E+02	0.11494
238	2	8.713	3.001	26.015	0.494	5.299	1.18E+03	0.15799
239	2	8.713	2.751	26.015	0.434	4.857	1.51E+03	0.19863
240	2	8.713	2.452	26.015	0.368	4.321	2.02E+03	0.25821
241	2	8.713	2.272	26.015	0.332	4.011	2.18E+03	0.27045
242	2	8.713	2.139	26.015	0.306	3.777	2.47E+03	0.3005
243	2	8.713	2.002	26.015	0.281	3.535	2.45E+03	0.28898
244	2	8.713	1.88	26.015	0.259	3.319	2.69E+03	0.30872
245	2	8.713	1.753	26.015	0.237	3.095	2.91E+03	0.32367
246	2	8.713	1.501	26.015	0.196	2.65	3.31E+03	0.33962
247	2	8.713	1.252	26.015	0.158	2.211	4.18E+03	0.3869
248	2	11.9	4.499	26.015	0.781	10.849	5.66E+01	0.01924
249	2	11.9	4.249	26.015	0.714	10.246	1.07E+02	0.03616
250	2	11.9	4	26.015	0.651	9.646	1.76E+02	0.05878
251	2	11.9	3.75	26.015	0.591	9.043	2.69E+02	0.08855
252	2	11.9	3.5	26.015	0.535	8.44	3.85E+02	0.12389
253	2	11.9	3.25	26.015	0.483	7.837	5.04E+02	0.15786
254	2	11.9	3	26.015	0.433	7.234	6.18E+02	0.18704
255	2	11.9	2.75	26.015	0.386	6.631	7.91E+02	0.2299
256	2	11.9	2.5	26.015	0.342	6.029	9.19E+02	0.25469
257	2	11.9	2.183	26.015	0.288	5.249	1.11E+03	0.28405
258	2	11.9	2	26.015	0.26	4.823	1.24E+03	0.30227
259	2	11.9	1.67	26.015	0.21	4.027	1.64E+03	0.36028
260	2	15.022	5.25	26.015	0.872	15.981	5.70E+00	0.00383
261	2	15.022	5	26.015	0.809	15.22	1.23E+01	0.0082
262	2	15.022	4.75	26.015	0.75	14.459	3.39E+01	0.02221
263	2	15.022	4.5	26.015	0.694	13.698	5.93E+01	0.03823
264	2	15.022	4.25	26.015	0.64	12.937	9.26E+01	0.05849
265	2	15.022	4	26.015	0.589	12.176	1.28E+02	0.07889
266	2	15.022	3.75	26.015	0.54	11.415	1.89E+02	0.11275
267	2	15.022	3.5	26.015	0.493	10.654	2.37E+02	0.13655
268	2	15.022	3.25	26.015	0.448	9.893	3.02E+02	0.16737
269	2	15.022	3	26.015	0.405	9.132	3.82E+02	0.20312

Table A.1: SLAC Proton DIS Data

I	J	E_0	E'	θ	x	Q^2	σ	F_2
270	2	15.022	2.75	26.015	0.364	8.371	4.64E+02	0.23352
271	2	15.022	2.463	26.015	0.317	7.48	5.58E+02	0.26205
272	2	15.022	2.25	26.015	0.286	6.849	6.29E+02	0.27877
273	2	15.022	2	26.015	0.249	6.088	6.57E+02	0.26916
274	2	15.022	1.75	26.015	0.214	5.327	8.87E+02	0.3327
275	2	18.028	5.496	26.015	0.854	20.078	3.98E+00	0.00443
276	2	18.028	5.246	26.015	0.799	19.165	9.99E+00	0.01094
277	2	18.028	4.996	26.015	0.746	18.252	1.89E+01	0.02032
278	2	18.028	4.625	26.015	0.67	16.857	3.84E+01	0.03974
279	2	18.028	4.244	26.015	0.599	15.504	7.59E+01	0.0755
280	2	18.028	3.994	26.015	0.554	14.591	9.98E+01	0.09612
281	2	18.028	3.745	26.015	0.51	13.681	1.30E+02	0.12024
282	2	18.028	3.493	26.015	0.468	12.761	1.71E+02	0.15229
283	2	18.028	3.247	26.015	0.428	11.862	2.25E+02	0.19112
284	2	18.028	3	26.015	0.389	10.96	2.76E+02	0.22259
285	2	18.028	2.743	26.015	0.349	10.021	3.14E+02	0.23873
286	2	18.028	2.5	26.015	0.313	9.133	3.54E+02	0.25218
287	2	18.028	2.25	26.015	0.278	8.22	4.27E+02	0.28152
288	2	18.028	2	26.015	0.243	7.306	4.52E+02	0.27444
289	2	4.502	1.8	34.009	0.547	2.772	2.82E+03	0.15086
290	2	4.502	1.6	34.009	0.453	2.464	4.01E+03	0.2155
291	2	4.502	1.4	34.009	0.37	2.156	4.81E+03	0.25417
292	2	4.502	1.2	34.009	0.298	1.848	5.91E+03	0.30018
293	2	4.502	1	34.009	0.234	1.54	6.95E+03	0.3296
294	2	5.808	2.25	34.009	0.67	4.471	6.66E+02	0.07533
295	2	5.808	2	34.009	0.556	3.974	1.15E+03	0.12876
296	2	5.808	1.75	34.009	0.457	3.477	1.79E+03	0.19547
297	2	5.808	1.5	34.009	0.369	2.98	2.48E+03	0.25876
298	2	5.808	1.25	34.009	0.29	2.484	3.12E+03	0.30305
299	2	5.808	1	34.009	0.22	1.987	3.95E+03	0.34351
300	2	7.912	2.75	34.009	0.768	7.443	9.75E+01	0.02715
301	2	7.912	2.5	34.009	0.666	6.767	2.20E+02	0.06006
302	2	7.912	2.25	34.009	0.573	6.09	4.13E+02	0.10931
303	2	7.912	2	34.009	0.488	5.413	6.44E+02	0.16317
304	2	7.912	1.751	34.009	0.41	4.739	9.21E+02	0.21942
305	2	7.912	1.48	34.009	0.332	4.006	1.27E+03	0.27786
306	2	10.014	3	34.009	0.781	10.277	3.49E+01	0.01878
307	2	10.014	2.75	34.009	0.691	9.421	7.53E+01	0.03921
308	2	10.014	2.5	34.009	0.607	8.565	1.49E+02	0.07442
309	2	10.014	2.25	34.009	0.529	7.708	2.77E+02	0.13197
310	2	10.014	2	34.009	0.456	6.852	4.10E+02	0.18331
311	2	10.014	1.75	34.009	0.387	5.995	5.67E+02	0.23404
312	2	10.014	1.5	34.009	0.322	5.139	7.84E+02	0.29444
313	2	12.518	3.25	34.009	0.8	13.918	8.65E+00	0.00851
314	2	12.518	3	34.009	0.719	12.847	3.15E+01	0.02981
315	2	12.518	2.75	34.009	0.642	11.777	6.27E+01	0.05677
316	2	12.518	2.5	34.009	0.569	10.706	1.10E+02	0.09432
317	2	12.518	2.25	34.009	0.5	9.635	1.85E+02	0.14876
318	2	12.518	2	34.009	0.434	8.565	2.25E+02	0.16805
319	2	12.518	1.75	34.009	0.371	7.494	3.67E+02	0.2501
320	2	15.012	3.25	34.009	0.756	16.691	1.32E+01	0.02022
321	2	15.012	3	34.009	0.683	15.407	2.64E+01	0.03869
322	2	15.012	2.75	34.009	0.614	14.123	4.64E+01	0.06446
323	2	15.012	2.5	34.009	0.547	12.839	8.16E+01	0.10629

Table A.1: SLAC Proton DIS Data

I	J	E_0	E'	θ	x	Q^2	σ	F_2
324	2	15.012	2.25	34.009	0.482	11.555	1.11E+02	0.13464
325	2	15.012	2	34.009	0.421	10.271	1.54E+02	0.17123
326	3	20.005	17.836	4	0.427	1.738	9.42E+05	0.23255
327	3	20.005	17.586	4	0.377	1.714	9.41E+05	0.25877
328	3	20.005	17.254	4	0.326	1.681	9.26E+05	0.28913
329	3	20.005	16.82	4	0.274	1.638	8.77E+05	0.31591
330	3	20.005	16.245	4	0.224	1.582	7.82E+05	0.33095
331	3	20.005	15.451	4	0.176	1.504	6.73E+05	0.34176
332	3	20.005	14.536	4	0.138	1.415	5.73E+05	0.34502
333	3	20.005	13.674	4	0.112	1.331	4.95E+05	0.33988
334	3	20.005	13.128	4	0.099	1.279	4.58E+05	0.33769
335	3	20.005	12.339	4	0.084	1.203	4.14E+05	0.3336
336	3	20.005	11.358	4	0.068	1.107	3.76E+05	0.33132
337	3	18.01	16.003	4	0.373	1.404	1.41E+06	0.26146
338	3	18.01	15.731	4	0.323	1.38	1.40E+06	0.29386
339	3	18.01	15.364	4	0.271	1.347	1.30E+06	0.31606
340	3	18.01	14.864	4	0.221	1.303	1.16E+06	0.33302
341	3	18.01	14.266	4	0.178	1.249	9.91E+05	0.33724
342	3	18.01	13.072	4	0.124	1.147	7.46E+05	0.32884
343	3	18.01	12.276	4	0.1	1.077	6.59E+05	0.33206
344	3	18.01	11.417	4	0.081	1.002	5.68E+05	0.32187
345	3	18.01	10.494	4	0.065	0.921	5.05E+05	0.31655
346	3	16	14.133	4	0.314	1.101	2.11E+06	0.28637
347	3	16	13.84	4	0.266	1.078	2.00E+06	0.31388
348	3	16	13.445	4	0.218	1.047	1.74E+06	0.32237
349	3	16	12.826	4	0.168	0.998	1.45E+06	0.33035
350	3	16	12.032	4	0.126	0.937	1.17E+06	0.32808
351	3	16	11.208	4	0.097	0.874	9.42E+05	0.31517
352	3	16	10.408	4	0.077	0.811	7.96E+05	0.30415
353	3	13	11.329	4	0.229	0.717	3.77E+06	0.30204
354	3	13	10.945	4	0.179	0.692	3.18E+06	0.31268
355	3	13	10.367	4	0.133	0.655	2.46E+06	0.30644
356	3	13	9.755	4	0.101	0.617	1.94E+06	0.29387
357	3	13	9.182	4	0.081	0.581	1.60E+06	0.28106
358	4	18.001	9.607	14.991	0.747	11.769	2.03E+02	0.02557
359	4	18.001	9.337	14.991	0.703	11.437	2.83E+02	0.03615
360	4	18.001	8.962	14.991	0.647	10.977	4.47E+02	0.05833
361	4	18.001	8.65	14.991	0.604	10.594	5.85E+02	0.07734
362	4	18.001	8.254	14.991	0.553	10.106	7.79E+02	0.10451
363	4	18.001	7.79	14.991	0.498	9.537	1.01E+03	0.13711
364	4	18.001	7.343	14.991	0.449	8.989	1.31E+03	0.17838
365	4	18.001	6.905	14.991	0.406	8.453	1.53E+03	0.20824
366	4	18.001	6.494	14.991	0.368	7.95	1.74E+03	0.23555
367	4	18.001	6.098	14.991	0.334	7.465	1.90E+03	0.25587
368	4	18.001	5.832	14.991	0.313	7.144	2.02E+03	0.26914
369	4	15.004	8.858	14.991	0.784	9.045	3.14E+02	0.02116
370	4	15.004	8.635	14.991	0.738	8.817	4.47E+02	0.03081
371	4	15.004	8.355	14.991	0.684	8.53	6.84E+02	0.04856
372	4	15.004	8.039	14.991	0.628	8.207	9.73E+02	0.07092
373	4	15.004	7.697	14.991	0.573	7.857	1.35E+03	0.10116
374	4	15.004	7.391	14.991	0.528	7.544	1.63E+03	0.12412
375	4	15.004	7.002	14.991	0.476	7.145	2.08E+03	0.16197
376	4	15.004	6.603	14.991	0.427	6.737	2.44E+03	0.19193
377	4	15.004	6.195	14.991	0.382	6.321	2.77E+03	0.22

58

Table A.1: SLAC Proton DIS Data

I	J	E_0	E'	θ	x	Q^2	σ	F_2
378	4	15.004	5.894	14.991	0.352	6.017	3.10E+03	0.24726
379	4	12.497	7.703	14.991	0.728	6.551	1.12E+03	0.04162
380	4	12.497	7.494	14.991	0.679	6.373	1.45E+03	0.05577
381	4	12.497	7.247	14.991	0.625	6.162	2.10E+03	0.08358
382	4	12.497	6.953	14.991	0.568	5.911	2.76E+03	0.11355
383	4	12.497	6.639	14.991	0.513	5.643	3.36E+03	0.14318
384	4	12.497	6.297	14.991	0.46	5.352	4.07E+03	0.17844
385	4	12.497	5.981	14.991	0.416	5.084	4.82E+03	0.21626
386	4	12.497	5.733	14.991	0.384	4.877	5.10E+03	0.2323
387	4	9.999	6.398	14.991	0.644	4.354	4.54E+03	0.08284
388	4	9.999	6.215	14.991	0.595	4.228	5.64E+03	0.10693
389	4	9.999	5.996	14.991	0.543	4.079	6.92E+03	0.13689
390	4	9.999	5.818	14.991	0.505	3.959	7.70E+03	0.15706
391	4	20.001	8.56	18.996	0.868	18.643	7.95E+00	0.00374
392	4	20.001	8.3	18.996	0.823	18.075	1.68E+01	0.00791
393	4	20.001	7.978	18.996	0.77	17.373	3.15E+01	0.01483
394	4	20.001	7.655	18.996	0.719	16.668	5.70E+01	0.02685
395	4	20.001	7.309	18.996	0.668	15.914	9.17E+01	0.0429
396	4	20.001	6.94	18.996	0.616	15.106	1.41E+02	0.06542
397	4	20.001	6.544	18.996	0.564	14.243	2.01E+02	0.09219
398	4	20.001	6.158	18.996	0.516	13.401	2.66E+02	0.11934
399	4	20.001	5.776	18.996	0.471	12.571	3.46E+02	0.15233
400	4	20.001	5.465	18.996	0.436	11.896	3.95E+02	0.17013
401	4	20.001	5.237	18.996	0.412	11.407	5.22E+02	0.22018
402	4	18.018	8.107	18.996	0.855	15.906	1.59E+01	0.00544
403	4	18.018	7.86	18.996	0.809	15.42	3.01E+01	0.01033
404	4	18.018	7.542	18.996	0.753	14.796	5.79E+01	0.02
405	4	18.018	7.237	18.996	0.702	14.195	9.54E+01	0.03295
406	4	18.018	6.91	18.996	0.65	13.553	1.47E+02	0.05076
407	4	18.018	6.561	18.996	0.598	12.866	2.12E+02	0.07268
408	4	18.018	6.191	18.996	0.547	12.138	3.06E+02	0.10386
409	4	18.018	5.747	18.996	0.489	11.268	4.20E+02	0.14013
410	4	18.018	5.479	18.996	0.457	10.743	4.93E+02	0.16191
411	4	18.018	5.27	18.996	0.432	10.342	5.63E+02	0.18302
412	4	15.004	7.274	18.996	0.819	11.884	5.37E+01	0.01042
413	4	15.004	7.061	18.996	0.774	11.536	9.56E+01	0.01871
414	4	15.004	6.807	18.996	0.723	11.12	1.60E+02	0.03161
415	4	15.004	6.522	18.996	0.669	10.653	2.40E+02	0.04779
416	4	15.004	6.221	18.996	0.616	10.161	3.59E+02	0.07192
417	4	15.004	5.91	18.996	0.566	9.651	4.93E+02	0.09895
418	4	15.004	5.577	18.996	0.515	9.106	6.65E+02	0.13289
419	4	15.004	5.322	18.996	0.479	8.695	7.34E+02	0.14595
420	4	12.498	6.47	18.996	0.778	8.805	2.00E+02	0.02185
421	4	12.498	6.254	18.996	0.726	8.51	3.25E+02	0.03608
422	4	12.498	6.017	18.996	0.673	8.187	4.85E+02	0.05467
423	4	12.498	5.764	18.996	0.621	7.843	6.85E+02	0.07842
424	4	12.498	5.513	18.996	0.572	7.501	8.82E+02	0.10199
425	4	12.498	5.301	18.996	0.534	7.215	1.14E+03	0.13238
426	4	18.018	5.601	25.993	0.876	20.41	2.70E+00	0.00301
427	4	18.018	5.407	25.993	0.832	19.7	5.66E+00	0.00623
428	4	18.018	5.161	25.993	0.779	18.803	1.23E+01	0.01332
429	4	18.018	4.932	25.993	0.732	17.964	2.11E+01	0.02241
430	4	18.018	4.646	25.993	0.674	16.911	3.65E+01	0.03776
431	4	18.018	4.356	25.993	0.618	15.85	6.20E+01	0.06205

Table A.1: SLAC Proton DIS Data

I	J	E_0	E'	θ	x	Q^2	σ	F_2
432	4	18.018	4.069	25.993	0.566	14.812	9.18E+01	0.08884
433	4	18.018	3.822	25.993	0.522	13.908	1.19E+02	0.11082
434	4	18.018	3.578	25.993	0.48	13.012	1.62E+02	0.14545
435	4	18.018	3.451	25.993	0.46	12.579	1.73E+02	0.15203
436	4	15.005	5.175	25.993	0.851	15.704	8.54E+00	0.00568
437	4	15.005	4.976	25.993	0.802	15.097	1.75E+01	0.01154
438	4	15.005	4.763	25.993	0.752	14.45	3.03E+01	0.01977
439	4	15.005	4.514	25.993	0.696	13.693	5.61E+01	0.03592
440	4	15.005	4.261	25.993	0.641	12.928	8.77E+01	0.05507
441	4	15.005	3.997	25.993	0.587	12.127	1.33E+02	0.08112
442	4	15.005	3.748	25.993	0.538	11.373	1.86E+02	0.1103
443	4	15.005	3.519	25.993	0.496	10.681	2.46E+02	0.14175
444	4	11.884	4.606	25.993	0.81	11.069	3.68E+01	0.01245
445	4	11.884	4.46	25.993	0.769	10.72	6.13E+01	0.02065
446	4	11.884	4.245	25.993	0.712	10.203	1.09E+02	0.03661
447	4	11.884	3.996	25.993	0.649	9.603	1.74E+02	0.05763
448	4	11.884	3.747	25.993	0.59	9.005	2.63E+02	0.08576
449	4	11.884	3.498	25.993	0.534	8.407	3.72E+02	0.11887
450	4	8.7	3.823	25.993	0.735	6.726	2.85E+02	0.03783
451	4	8.7	3.676	25.993	0.686	6.467	4.09E+02	0.05471
452	4	8.7	3.518	25.993	0.636	6.189	5.54E+02	0.07437
453	4	8.7	3.392	25.993	0.599	5.97	6.30E+02	0.08462
454	4	15.004	3.5	33.992	0.831	17.948	4.28E+00	0.00682
455	4	15.004	3.25	33.992	0.756	16.666	1.18E+01	0.01803
456	4	15.004	3	33.992	0.683	15.384	2.56E+01	0.03732
457	4	15.004	2.75	33.992	0.613	14.102	4.79E+01	0.0663
458	4	15.004	2.5	33.992	0.546	12.82	8.09E+01	0.10531
459	4	12.499	3.25	33.992	0.8	13.883	1.31E+01	0.01282
460	4	12.499	3	33.992	0.719	12.815	3.18E+01	0.02998
461	4	12.499	2.75	33.992	0.642	11.747	6.54E+01	0.059
462	4	12.499	2.5	33.992	0.569	10.679	1.09E+02	0.09301
463	4	9.999	3	33.992	0.781	10.252	3.53E+01	0.01888
464	4	9.999	2.75	33.992	0.691	9.398	8.08E+01	0.04192
465	4	9.999	2.5	33.992	0.607	8.543	1.73E+02	0.08635
466	4	9.999	2.25	33.992	0.529	7.689	2.55E+02	0.12065
467	5	19.505	1.557	60	0.9	30.314	5.99E-02	0.00077
468	5	19.505	1.47	60	0.845	28.6	2.70E-01	0.00331
469	5	19.505	1.374	60	0.785	26.693	8.02E-01	0.00924
470	5	19.505	1.278	60	0.727	24.851	1.80E+00	0.01963
471	5	19.505	1.181	60	0.667	22.931	3.77E+00	0.03799
472	5	19.505	1.099	60	0.619	21.388	6.20E+00	0.05925
473	5	19.505	1.039	60	0.584	20.23	8.23E+00	0.07422
474	5	16.002	1.506	59.999	0.885	24.068	2.24E-01	0.00184
475	5	16.002	1.433	59.999	0.838	22.899	5.94E-01	0.00466
476	5	16.002	1.352	59.999	0.785	21.584	1.43E+00	0.0108
477	5	16.002	1.259	59.999	0.726	20.078	2.99E+00	0.02124
478	5	16.002	1.175	59.999	0.675	18.773	5.21E+00	0.03497
479	5	16.002	1.108	59.999	0.633	17.693	8.48E+00	0.05403
480	5	16.002	1.044	59.999	0.595	16.706	1.39E+01	0.08354
481	5	13.29	1.46	60	0.873	19.382	5.64E-01	0.00301
482	5	13.29	1.39	60	0.826	18.442	1.26E+00	0.00649
483	5	13.29	1.311	60	0.773	17.384	2.66E+00	0.01302
484	5	13.29	1.236	60	0.725	16.393	4.70E+00	0.02206
485	5	13.29	1.158	60	0.674	15.345	8.33E+00	0.03708

Table A.1: SLAC Proton DIS Data

I	J	E_0	E'	θ	x	Q^2	σ	F_2
486	5	13.29	1.083	60	0.626	14.348	1.53E+01	0.06419
487	5	13.29	1.009	60	0.58	13.376	2.08E+01	0.08283
488	5	13.29	0.949	60	0.543	12.569	2.95E+01	0.11197
489	5	13.29	0.899	60	0.514	11.945	3.91E+01	0.14203
490	5	10.4	1.375	60	0.844	14.288	2.03E+00	0.00607
491	5	10.4	1.303	60	0.792	13.527	4.72E+00	0.01355
492	5	10.4	1.222	60	0.736	12.684	9.45E+00	0.02584
493	5	10.4	1.141	60	0.681	11.839	1.62E+01	0.04202
494	5	10.4	1.067	60	0.632	11.076	2.47E+01	0.0611
495	5	10.4	1.003	60	0.59	10.41	3.52E+01	0.08325
496	5	10.4	0.934	60	0.545	9.682	4.94E+01	0.1104
497	5	10.4	0.875	60	0.509	9.094	6.02E+01	0.1273
498	5	6.5	1.168	59.999	0.758	7.587	3.10E+01	0.02856
499	5	6.5	1.102	59.999	0.706	7.15	5.02E+01	0.04467
500	5	6.5	1.034	59.999	0.654	6.71	7.73E+01	0.06627
501	5	6.5	0.974	59.999	0.609	6.317	1.08E+02	0.08895
502	5	6.5	0.918	59.999	0.569	5.963	1.49E+02	0.11829
503	5	6.5	0.853	59.999	0.522	5.534	1.79E+02	0.13559
504	5	6.5	0.798	59.999	0.483	5.173	2.44E+02	0.17661
505	5	6.5	0.739	59.999	0.444	4.796	3.36E+02	0.23057
506	5	6.5	0.694	59.999	0.413	4.503	4.02E+02	0.26369
507	5	6.5	0.658	59.999	0.39	4.277	3.29E+02	0.20854
508	5	19.5	1.701	49.994	0.708	23.662	3.54E+00	0.02551
509	5	19.5	1.594	49.994	0.658	22.121	6.05E+00	0.0411
510	5	19.5	1.483	49.994	0.61	20.616	8.96E+00	0.05757
511	5	19.5	1.386	49.994	0.566	19.255	1.32E+01	0.07966
512	5	19.5	1.289	49.994	0.523	17.89	1.76E+01	0.09986
513	5	19.5	1.181	49.994	0.476	16.364	2.76E+01	0.14585
514	5	19.5	1.083	49.994	0.434	15.014	3.42E+01	0.16747
515	5	16	1.712	49.995	0.729	19.539	3.84E+00	0.01815
516	5	16	1.603	49.995	0.676	18.26	8.11E+00	0.03638
517	5	16	1.483	49.995	0.621	16.913	1.30E+01	0.05468
518	5	16	1.374	49.995	0.571	15.667	2.00E+01	0.0795
519	5	16	1.27	49.995	0.523	14.464	2.94E+01	0.10953
520	5	16	1.156	49.995	0.472	13.136	4.04E+01	0.13882
521	5	16	1.059	49.995	0.432	12.103	5.17E+01	0.16567
522	5	16	0.997	49.995	0.405	11.394	5.85E+01	0.17816
523	5	16	0.933	49.995	0.377	10.663	4.32E+01	0.12488
524	5	13.5	1.702	49.998	0.74	16.389	6.58E+00	0.0214
525	5	13.5	1.603	49.998	0.692	15.438	1.13E+01	0.03512
526	5	13.5	1.508	49.998	0.645	14.517	1.76E+01	0.0521
527	5	13.5	1.403	49.998	0.595	13.496	2.42E+01	0.06803
528	5	13.5	1.303	49.998	0.548	12.541	3.84E+01	0.10184
529	5	13.5	1.208	49.998	0.504	11.619	5.07E+01	0.12628
530	5	13.5	1.127	49.998	0.466	10.825	6.38E+01	0.15047
531	5	13.5	1.041	49.998	0.429	10.024	7.85E+01	0.17316
532	5	13.5	0.981	49.998	0.402	9.444	8.11E+01	0.17153
533	5	7	1.524	49.998	0.741	7.619	5.40E+01	0.03639
534	5	7	1.449	49.998	0.695	7.241	6.67E+01	0.04369
535	5	7	1.366	49.998	0.645	6.819	9.80E+01	0.06221
536	5	7	1.265	49.998	0.587	6.316	1.59E+02	0.0964
537	5	7	1.182	49.998	0.541	5.906	2.10E+02	0.12261
538	5	7	1.11	49.998	0.502	5.545	2.72E+02	0.15323
539	5	7	1.035	49.998	0.462	5.168	3.27E+02	0.17607

61

Table A.1: SLAC Proton DIS Data

I	J	E_0	E'	θ	x	Q^2	σ	F_2
540	5	7	0.959	49.998	0.422	4.784	4.54E+02	0.23191
541	5	7	0.877	49.998	0.381	4.374	5.37E+02	0.25957
542	5	7	0.803	49.998	0.344	4.003	5.90E+02	0.26768
543	5	7	0.729	49.998	0.31	3.645	1.06E+03	0.45255
544	6	13.5	11.358	6	0.418	1.68	4.37E+05	0.24332
545	6	13.5	11.094	6	0.363	1.64	4.46E+05	0.27827
546	6	13.5	10.862	6	0.324	1.606	4.55E+05	0.31012
547	6	13.5	9.852	6	0.213	1.457	3.65E+05	0.33726
548	6	13.5	8.494	6	0.134	1.256	2.85E+05	0.34708
549	6	13.5	6.89	6	0.082	1.019	2.24E+05	0.33099
550	6	13.5	5.997	6	0.063	0.887	2.02E+05	0.31594
551	6	16	13.552	6	0.517	2.375	2.08E+05	0.18571
552	6	16	13.235	6	0.447	2.32	2.11E+05	0.21233
553	6	16	13.001	6	0.405	2.279	2.28E+05	0.24854
554	6	16	12.585	6	0.344	2.206	2.10E+05	0.25858
555	6	16	12.006	6	0.281	2.105	2.19E+05	0.31351
556	6	16	11.367	6	0.229	1.993	2.03E+05	0.33275
557	6	16	10.664	6	0.187	1.869	1.85E+05	0.34294
558	6	16	9.904	6	0.152	1.736	1.68E+05	0.34733
559	6	16	9.081	6	0.123	1.592	1.53E+05	0.34767
560	6	16	7.256	6	0.078	1.272	1.33E+05	0.34498
561	6	19.5	16.381	6	0.598	3.499	6.80E+04	0.11508
562	6	19.5	16.148	6	0.548	3.45	7.65E+04	0.13873
563	6	19.5	15.845	6	0.493	3.384	9.02E+04	0.17763
564	6	19.5	15.495	6	0.44	3.31	9.74E+04	0.20953
565	6	19.5	14.926	6	0.372	3.189	1.04E+05	0.25441
566	6	19.5	14.3	6	0.313	3.055	1.04E+05	0.28573
567	6	19.5	13.488	6	0.255	2.877	1.02E+05	0.31998
568	6	19.5	12.059	6	0.185	2.576	9.07E+04	0.33889
569	6	19.5	11.192	6	0.153	2.391	8.86E+04	0.35985
570	6	19.5	10.266	6	0.127	2.193	8.05E+04	0.3505
571	6	19.5	9.28	6	0.103	1.983	7.65E+04	0.35163
572	6	19.5	8.234	6	0.083	1.759	7.27E+04	0.34599
573	6	19.5	7.128	6	0.066	1.523	7.55E+04	0.36294
574	6	19.5	13.83	10	0.77	8.193	1.08E+03	0.0243
575	6	19.5	13.547	10	0.718	8.025	1.57E+03	0.03695
576	6	19.5	13.253	10	0.67	7.851	2.25E+03	0.05515
577	6	19.5	12.85	10	0.61	7.614	3.13E+03	0.08094
578	6	19.5	12.228	10	0.531	7.245	4.53E+03	0.12564
579	6	19.5	11.15	10	0.422	6.606	6.33E+03	0.1943
580	6	19.5	10.246	10	0.35	6.071	7.65E+03	0.24979
581	6	19.5	8.816	10	0.261	5.223	8.64E+03	0.30112
582	6	19.5	6.893	10	0.173	4.084	9.77E+03	0.34753
583	6	13.3	8.093	15	0.751	7.334	6.84E+02	0.03119
584	6	13.3	7.888	15	0.704	7.148	9.71E+02	0.04556
585	6	13.3	7.68	15	0.66	6.96	1.32E+03	0.06368
586	6	13.3	7.455	15	0.616	6.757	1.67E+03	0.08238
587	6	13.3	6.996	15	0.536	6.341	2.43E+03	0.12549
588	6	13.3	6.231	15	0.426	5.648	3.71E+03	0.20198
589	6	13.3	5.616	15	0.353	5.09	4.57E+03	0.25482
590	6	13.3	4.691	15	0.263	4.252	5.49E+03	0.3069
591	6	13.3	3.528	15	0.174	3.198	7.15E+03	0.3754
592	6	16	9.279	15	0.802	10.115	1.85E+02	0.01542
593	6	16	9.066	15	0.76	9.883	2.88E+02	0.02445

62

Table A.1: SLAC Proton DIS Data

I	J	E_0	E'	θ	x	Q^2	σ	F_2
594	6	16	8.684	15	0.69	9.468	5.01E+02	0.04402
595	6	16	8.173	15	0.607	8.912	8.71E+02	0.07926
596	6	16	7.625	15	0.529	8.314	1.32E+03	0.12415
597	6	16	6.724	15	0.421	7.332	2.06E+03	0.19824
598	6	16	6.014	15	0.35	6.558	2.64E+03	0.25508
599	6	16	4.967	15	0.262	5.416	3.31E+03	0.30938
600	6	16	3.685	15	0.174	4.018	4.10E+03	0.34706
601	6	19.5	10.466	15	0.82	13.907	6.07E+01	0.00968
602	6	19.5	10.264	15	0.787	13.638	9.25E+01	0.01494
603	6	19.5	10.022	15	0.749	13.318	1.37E+02	0.02247
604	6	19.5	9.514	15	0.675	12.643	2.66E+02	0.04457
605	6	19.5	8.913	15	0.596	11.844	4.57E+02	0.0781
606	6	19.5	8.261	15	0.521	10.978	7.01E+02	0.12138
607	6	19.5	7.206	15	0.415	9.576	1.14E+03	0.19724
608	6	19.5	4.652	15	0.222	6.182	2.15E+03	0.32562
609	6	19.5	3.805	15	0.172	5.056	2.59E+03	0.35515
610	6	6.5	3.925	18	0.517	2.497	1.49E+04	0.16592
611	6	6.5	3.826	18	0.485	2.434	1.60E+04	0.18355
612	6	6.5	3.603	18	0.421	2.292	1.90E+04	0.23059
613	6	6.5	3.188	18	0.326	2.028	2.17E+04	0.2859
614	6	6.5	2.714	18	0.243	1.727	2.38E+04	0.32888
615	6	6.5	2.193	18	0.173	1.395	2.31E+04	0.32028
616	6	10.4	5.924	18	0.718	6.029	9.18E+02	0.04398
617	6	10.4	5.683	18	0.653	5.782	1.39E+03	0.06894
618	6	10.4	5.525	18	0.615	5.624	1.77E+03	0.08939
619	6	10.4	5.209	18	0.544	5.303	2.44E+03	0.1271
620	6	10.4	4.61	18	0.432	4.693	3.74E+03	0.20332
621	6	10.4	4.133	18	0.358	4.208	4.77E+03	0.26355
622	6	10.4	3.426	18	0.267	3.488	5.70E+03	0.31132
623	6	10.4	2.552	18	0.176	2.598	6.96E+03	0.35162
624	6	13.3	7.067	18	0.786	9.198	1.79E+02	0.01881
625	6	13.3	6.897	18	0.747	8.978	2.64E+02	0.02809
626	6	13.3	6.631	18	0.69	8.633	4.18E+02	0.04538
627	6	13.3	6.22	18	0.609	8.098	7.23E+02	0.08041
628	6	13.3	5.777	18	0.533	7.521	1.09E+03	0.12318
629	6	13.3	5.047	18	0.424	6.571	1.74E+03	0.19856
630	6	13.3	4.483	18	0.353	5.836	2.24E+03	0.25168
631	6	13.3	3.666	18	0.264	4.773	2.91E+03	0.30994
632	6	13.3	2.688	18	0.176	3.5	3.79E+03	0.35453
633	6	16	7.922	18	0.818	12.405	5.54E+01	0.01046
634	6	16	7.641	18	0.763	11.963	1.09E+02	0.02075
635	6	16	7.174	18	0.678	11.236	2.26E+02	0.04402
636	6	16	6.688	18	0.599	10.475	4.00E+02	0.07859
637	6	16	6.165	18	0.523	9.656	6.19E+02	0.12169
638	6	16	5.331	18	0.417	8.349	1.01E+03	0.19431
639	6	16	4.696	18	0.347	7.355	1.36E+03	0.25248
640	6	16	3.793	18	0.259	5.941	1.82E+03	0.31144
641	6	16	2.739	18	0.172	4.29	2.35E+03	0.34086
642	6	19.5	8.933	18	0.86	17.049	1.27E+01	0.00445
643	6	19.5	8.627	18	0.807	16.463	2.91E+01	0.01024
644	6	19.5	8.228	18	0.742	15.706	6.24E+01	0.02208
645	6	19.5	7.742	18	0.67	14.778	1.22E+02	0.0432
646	6	19.5	7.174	18	0.592	13.694	2.16E+02	0.07591
647	6	19.5	6.573	18	0.517	12.547	3.49E+02	0.12107

63

Table A.1: SLAC Proton DIS Data

I	J	E_0	E'	θ	x	Q^2	σ	F_2
648	6	19.5	5.628	18	0.413	10.743	6.05E+02	0.20035
649	6	19.5	4.92	18	0.343	9.391	7.71E+02	0.24104
650	6	19.5	3.93	18	0.257	7.502	1.09E+03	0.30261
651	6	19.5	2.803	18	0.171	5.35	1.53E+03	0.34961
652	6	19.5	7.693	20.6	0.866	19.18	5.69E+00	0.00341
653	6	19.5	7.383	20.6	0.809	18.4	1.44E+01	0.00863
654	6	19.5	6.977	20.6	0.74	17.398	3.46E+01	0.02053
655	6	19.5	6.522	20.6	0.668	16.264	7.02E+01	0.04106
656	6	19.5	6.001	20.6	0.591	14.965	1.29E+02	0.07371
657	6	19.5	5.456	20.6	0.516	13.606	2.20E+02	0.12156
658	6	19.5	4.618	20.6	0.412	11.516	3.70E+02	0.18946
659	6	19.5	4.002	20.6	0.343	9.98	5.10E+02	0.24185
660	6	19.5	3.16	20.6	0.257	7.88	7.29E+02	0.29877
661	6	19.5	2.225	20.6	0.171	5.548	9.87E+02	0.32097

Appendix B
SLAC Deuteron DIS Data

This appendix contains the final cross sections and the F_2 values extracted from them from the analysis of the SLAC deep inelastic scattering data for the deuteron. The data is from Files E.3 and E.5 of the Ph.D. thesis for Stanford University, written by L.W. Whitlow entitled:

Deep Inelastic Structure Functions
from Electron Scattering on Hydrogen,
Deuterium, and Iron at 0.6 GeV² < Q^2 < 30.0 GeV²,

document number SLAC-357, dated March 1990.

In the table:

I   =   counting index
J   =   code for SLAC experiment number:
        1 = E49a, 2 = E49b, 3 = E61, 4 = E87,
        5 = E89a, 6 = E89b, 7 = E139, 8 = E140
E_0 =   incident beam energy (GeV)
E' =   scattered electron energy (GeV)
θ =   scattering angle (degrees)
x =   Bjorken scaling variable (momentum fraction)
Q^2 =   4-momentum transfer squared (GeV²)
σ =   final measured cross section (pb/srGeV)
F_2 =   final extracted structure function

Details concerning the collection, analysis, preparation and error treatment of this data are in the reference given above.

Table B.1: SLAC Deuteron DIS Data

I	J	E_0	E'	θ	x	Q^2	σ	F_2
1	1	10.027	8.273	5.988	0.275	0.905	1.05E+06	0.26169
2	1	10.027	7.954	5.988	0.223	0.869	9.45E+05	0.27658
3	1	10.027	7.538	5.988	0.176	0.824	8.20E+05	0.28495
4	1	10.027	7.257	5.988	0.153	0.794	7.48E+05	0.28688
5	1	10.027	6.731	5.988	0.119	0.737	6.37E+05	0.28484
6	1	10.027	6.088	5.988	0.09	0.666	5.30E+05	0.27439
7	1	10.027	5.351	5.988	0.067	0.586	4.39E+05	0.25588
8	1	13.549	11.396	5.988	0.417	1.685	3.54E+05	0.19828
9	1	13.549	11.145	5.988	0.365	1.647	3.63E+05	0.22649
10	1	13.549	10.836	5.988	0.314	1.601	3.56E+05	0.24937
11	1	13.549	10.501	5.988	0.271	1.552	3.44E+05	0.26875
12	1	13.549	9.844	5.988	0.209	1.455	3.13E+05	0.29341
13	1	13.549	9.218	5.988	0.168	1.363	2.84E+05	0.30559
14	1	13.549	8.492	5.988	0.132	1.256	2.51E+05	0.30726
15	1	13.549	7.715	5.988	0.104	1.141	2.20E+05	0.30117
16	1	13.549	6.889	5.988	0.081	1.019	2.05E+05	0.30495
17	1	16.075	13.556	5.988	0.503	2.377	1.53E+05	0.14116
18	1	16.075	13.309	5.988	0.45	2.334	1.66E+05	0.16714
19	1	16.075	13.012	5.988	0.397	2.282	1.77E+05	0.19747
20	1	16.075	12.673	5.988	0.348	2.222	1.81E+05	0.22336
21	1	16.075	12.09	5.988	0.283	2.119	1.78E+05	0.25459
22	1	16.075	11.373	5.988	0.226	1.995	1.75E+05	0.29103
23	1	16.075	10.67	5.988	0.185	1.872	1.63E+05	0.30606
24	1	16.075	9.908	5.988	0.15	1.738	1.50E+05	0.3139
25	1	16.075	9.087	5.988	0.122	1.594	1.38E+05	0.31621
26	1	16.075	8.208	5.988	0.098	1.44	1.30E+05	0.32056
27	1	16.075	7.27	5.988	0.077	1.275	1.22E+05	0.31897
28	1	16.075	6.478	5.988	0.063	1.136	1.16E+05	0.30884
29	1	19.544	16.42	5.988	0.597	3.501	5.13E+04	0.08659
30	1	19.544	16.157	5.988	0.542	3.445	5.99E+04	0.10926
31	1	19.544	15.865	5.988	0.49	3.383	6.92E+04	0.13666
32	1	19.544	15.545	5.988	0.442	3.314	7.55E+04	0.16153
33	1	19.544	14.967	5.988	0.372	3.191	8.28E+04	0.20132
34	1	19.544	14.287	5.988	0.309	3.047	8.81E+04	0.2434
35	1	19.544	13.255	5.988	0.239	2.819	8.66E+04	0.28035
36	1	19.544	12.046	5.988	0.183	2.569	8.11E+04	0.30387
37	1	19.544	11.171	5.988	0.152	2.382	7.95E+04	0.3234
38	1	19.544	10.261	5.988	0.126	2.188	7.44E+04	0.32423
39	1	19.544	9.455	5.988	0.107	2.017	7.21E+04	0.32854
40	1	19.544	8.65	5.988	0.09	1.845	6.95E+04	0.32732
41	1	19.544	7.774	5.988	0.075	1.658	6.94E+04	0.33283
42	1	7.019	5.187	10	0.322	1.105	2.54E+05	0.24564
43	1	7.019	4.938	10	0.269	1.052	2.43E+05	0.26365
44	1	7.019	4.762	10	0.24	1.016	2.38E+05	0.27708
45	1	7.019	4.276	10	0.177	0.912	2.12E+05	0.29002
46	1	7.019	3.71	10	0.127	0.791	1.87E+05	0.29122
47	1	7.019	3.082	10	0.089	0.657	1.63E+05	0.27477
48	1	9.022	6.824	10	0.453	1.87	8.70E+04	0.16718
49	1	9.022	6.352	10	0.347	1.741	9.97E+04	0.22877
50	1	9.022	5.358	10	0.214	1.469	9.66E+04	0.28851
51	1	9.022	4.069	10	0.12	1.115	8.41E+04	0.29981
52	1	10.998	8.269	10	0.539	2.762	3.34E+04	0.11804
53	1	10.998	8.062	10	0.489	2.693	3.73E+04	0.14083

66

I	J	E_0	E'	θ	x	Q^2	σ	F_2
54	1	10.998	7.831	10	0.44	2.616	4.19E+04	0.16938
55	1	10.998	7.452	10	0.374	2.489	4.68E+04	0.2088
56	1	10.998	6.875	10	0.297	2.297	4.95E+04	0.25065
57	1	10.998	6.279	10	0.237	2.098	5.08E+04	0.28499
58	1	10.998	5.627	10	0.187	1.88	4.97E+04	0.30281
59	1	10.998	4.919	10	0.144	1.644	4.78E+04	0.30853
60	1	10.998	4.155	10	0.108	1.388	4.81E+04	0.3181
61	1	10.998	3.335	10	0.077	1.114	4.79E+04	0.30769
62	1	13.545	10.083	10	0.639	4.149	9.70E+03	0.06569
63	1	13.545	9.828	10	0.58	4.044	1.22E+04	0.08822
64	1	13.545	9.697	10	0.553	3.991	1.32E+04	0.0986
65	1	13.545	9.243	10	0.471	3.804	1.76E+04	0.14484
66	1	13.545	8.733	10	0.398	3.594	2.07E+04	0.1862
67	1	13.545	8.157	10	0.332	3.357	2.28E+04	0.2248
68	1	13.545	7.514	10	0.273	3.092	2.47E+04	0.26331
69	1	13.545	6.849	10	0.224	2.819	2.55E+04	0.28982
70	1	13.545	6.096	10	0.179	2.509	2.55E+04	0.3044
71	1	13.545	5.32	10	0.142	2.189	2.63E+04	0.32172
72	1	13.545	4.456	10	0.108	1.834	2.71E+04	0.32934
73	1	13.545	3.746	10	0.084	1.542	2.71E+04	0.31463
74	1	15.204	11.214	10	0.692	5.18	4.35E+03	0.04257
75	1	15.204	10.785	10	0.601	4.981	7.16E+03	0.07687
76	1	15.204	10.392	10	0.532	4.8	9.00E+03	0.10404
77	1	15.204	9.856	10	0.454	4.553	1.19E+04	0.15011
78	1	15.204	9.296	10	0.387	4.294	1.39E+04	0.18893
79	1	15.204	8.688	10	0.328	4.014	1.59E+04	0.23263
80	1	15.204	8.007	10	0.274	3.699	1.64E+04	0.25541
81	1	15.204	7.301	10	0.227	3.373	1.80E+04	0.29471
82	1	15.204	6.522	10	0.185	3.013	1.83E+04	0.3106
83	1	15.204	5.695	10	0.147	2.631	1.86E+04	0.32128
84	1	15.204	4.989	10	0.12	2.305	1.91E+04	0.32609
85	1	15.204	4.259	10	0.096	1.968	1.99E+04	0.32755
86	1	17.706	12.789	10	0.746	6.879	1.53E+03	0.02488
87	1	17.706	12.51	10	0.69	6.729	2.11E+03	0.03601
88	1	17.706	12.356	10	0.662	6.647	2.28E+03	0.03998
89	1	17.706	11.972	10	0.599	6.441	3.73E+03	0.06929
90	1	17.706	11.477	10	0.528	6.174	4.98E+03	0.09901
91	1	17.706	10.956	10	0.465	5.894	6.42E+03	0.13584
92	1	17.706	10.352	10	0.404	5.569	7.77E+03	0.17504
93	1	17.706	9.693	10	0.347	5.215	8.98E+03	0.21421
94	1	17.706	9.006	10	0.297	4.845	9.93E+03	0.24817
95	1	17.706	8.265	10	0.251	4.446	1.05E+04	0.27297
96	1	17.706	7.441	10	0.208	4.003	1.13E+04	0.30075
97	1	17.706	6.782	10	0.178	3.649	1.16E+04	0.31109
98	1	17.706	6.069	10	0.15	3.265	1.23E+04	0.32923
99	1	17.706	5.3	10	0.122	2.851	1.30E+04	0.33957
100	1	17.706	4.531	10	0.099	2.438	1.30E+04	0.32544
101	1	17.706	3.735	10	0.077	2.009	1.48E+04	0.33899
102	1	19.35	13.758	10	0.771	8.088	7.84E+02	0.01722
103	1	19.35	13.434	10	0.711	7.897	1.30E+03	0.03007
104	1	19.35	13.024	10	0.645	7.656	1.98E+03	0.0483
105	1	19.35	12.498	10	0.571	7.348	2.87E+03	0.07487
106	1	19.35	11.969	10	0.508	7.037	3.83E+03	0.10573
107	1	19.35	11.381	10	0.447	6.691	4.86E+03	0.14195

Table B.1: SLAC Deuteron DIS Data

I	J	E_0	E'	θ	x	Q^2	σ	F_2
108	1	19.35	10.763	10	0.393	6.328	5.66E+03	0.17369
109	1	19.35	10.087	10	0.341	5.931	6.62E+03	0.2125
110	1	19.35	9.352	10	0.293	5.498	7.15E+03	0.23816
111	1	19.35	8.558	10	0.248	5.032	7.90E+03	0.27059
112	1	19.35	7.882	10	0.215	4.634	8.43E+03	0.29281
113	1	19.35	7.176	10	0.185	4.219	9.54E+03	0.33241
114	1	19.35	6.47	10	0.157	3.804	9.12E+03	0.31528
115	1	19.35	5.706	10	0.131	3.355	9.97E+03	0.33682
116	1	19.35	4.912	10	0.107	2.888	1.05E+04	0.33953
117	1	19.35	4.059	10	0.083	2.386	1.15E+04	0.3402
118	2	4.504	2.653	18.02	0.337	1.171	6.20E+04	0.23992
119	2	4.504	2.497	18.02	0.293	1.102	6.30E+04	0.2578
120	2	4.504	2.363	18.02	0.26	1.043	6.35E+04	0.27062
121	2	4.504	2.25	18.02	0.235	0.994	6.42E+04	0.28224
122	2	4.504	2	18.02	0.188	0.884	6.30E+04	0.28951
123	2	4.504	1.75	18.02	0.15	0.773	6.18E+04	0.28935
124	2	4.504	1.5	18.02	0.118	0.663	6.01E+04	0.27724
125	2	6.509	3.94	18.02	0.522	2.515	1.20E+04	0.1343
126	2	6.509	3.826	18.02	0.485	2.443	1.26E+04	0.14551
127	2	6.509	3.552	18.02	0.408	2.266	1.52E+04	0.18854
128	2	6.509	3.325	18.02	0.355	2.121	1.71E+04	0.22154
129	2	6.509	3.042	18.02	0.298	1.941	1.85E+04	0.25071
130	2	6.509	2.715	18.02	0.243	1.732	2.05E+04	0.28623
131	2	6.509	2.438	18.02	0.203	1.553	2.07E+04	0.29074
132	2	6.509	2.239	18.02	0.178	1.428	2.14E+04	0.29943
133	2	6.509	2.058	18.02	0.157	1.312	2.17E+04	0.29968
134	2	6.509	1.5	18.02	0.102	0.958	2.47E+04	0.30479
135	2	10.392	5.998	18.02	0.742	6.115	6.30E+02	0.02987
136	2	10.392	5.75	18.02	0.673	5.862	9.12E+02	0.04484
137	2	10.392	5.498	18.02	0.61	5.605	1.31E+03	0.06624
138	2	10.392	5.25	18.02	0.555	5.352	1.75E+03	0.09108
139	2	10.392	4.899	18.02	0.484	4.986	2.36E+03	0.12631
140	2	10.392	4.421	18.02	0.402	4.501	3.28E+03	0.18
141	2	10.392	4.031	18.02	0.344	4.103	3.87E+03	0.21413
142	2	10.392	3.713	18.02	0.302	3.78	4.36E+03	0.24059
143	2	10.392	3.439	18.02	0.268	3.501	4.78E+03	0.26184
144	2	10.392	3.192	18.02	0.241	3.254	5.13E+03	0.27734
145	2	10.392	2.923	18.02	0.212	2.973	5.56E+03	0.2932
146	2	10.392	2.688	18.02	0.19	2.74	5.88E+03	0.3026
147	2	10.392	2.497	18.02	0.172	2.546	5.91E+03	0.29612
148	2	10.392	2.317	18.02	0.156	2.362	6.39E+03	0.31049
149	2	10.392	2.136	18.02	0.141	2.178	6.09E+03	0.28484
150	2	10.392	1.997	18.02	0.129	2.036	6.88E+03	0.31055
151	2	10.392	1.792	18.02	0.113	1.827	7.47E+03	0.31733
152	2	10.392	1.497	18.02	0.091	1.526	8.33E+03	0.31502
153	2	12.518	5.914	18.02	0.586	7.263	7.61E+02	0.07075
154	2	12.518	4.947	18.02	0.428	6.075	1.67E+03	0.15953
155	2	12.518	4.121	18.02	0.321	5.061	2.38E+03	0.22346
156	2	12.518	3.314	18.02	0.236	4.07	3.12E+03	0.27646
157	2	12.518	2.944	18.02	0.201	3.615	3.45E+03	0.29223
158	2	12.518	2.689	18.02	0.179	3.302	3.77E+03	0.30751
159	2	12.518	2.428	18.02	0.157	2.982	4.04E+03	0.31325
160	2	12.518	2.249	18.02	0.143	2.762	4.34E+03	0.3241
161	2	12.518	2.07	18.02	0.13	2.542	4.55E+03	0.32427

68

Table B.1: SLAC Deuteron DIS Data

I	J	E_0	E'	θ	x	Q^2	σ	F_2
162	2	12.518	1.886	18.02	0.116	2.316	4.80E+03	0.32237
163	2	12.518	1.746	18.02	0.106	2.144	5.10E+03	0.326
164	2	12.518	1.605	18.02	0.096	1.971	5.18E+03	0.31347
165	2	12.518	1.362	18.02	0.08	1.665	5.82E+03	0.31135
166	2	13.32	7	18.02	0.771	9.147	1.55E+02	0.01654
167	2	13.32	6.5	18.02	0.664	8.494	3.61E+02	0.03989
168	2	13.32	5.828	18.02	0.541	7.607	8.09E+02	0.09208
169	2	13.32	5.385	18.02	0.472	7.026	1.12E+03	0.12826
170	2	13.32	4.876	18.02	0.401	6.359	1.52E+03	0.17389
171	2	13.32	4.349	18.02	0.337	5.674	1.89E+03	0.21258
172	2	13.32	3.923	18.02	0.29	5.118	2.22E+03	0.24282
173	2	13.32	3.601	18.02	0.257	4.696	2.49E+03	0.26508
174	2	13.32	3.344	18.02	0.233	4.37	2.71E+03	0.28058
175	2	13.32	3.022	18.02	0.204	3.938	2.89E+03	0.28755
176	2	13.32	2.702	18.02	0.177	3.526	3.24E+03	0.30659
177	2	13.32	2.5	18.02	0.161	3.267	3.44E+03	0.31355
178	2	13.32	2.312	18.02	0.146	3.021	3.71E+03	0.32391
179	2	13.32	1.972	18.02	0.121	2.572	3.96E+03	0.3151
180	2	13.32	1.786	18.02	0.108	2.334	4.19E+03	0.31224
181	2	13.32	1.645	18.02	0.098	2.146	4.77E+03	0.33653
182	2	13.32	1.5	18.02	0.088	1.96	5.19E+03	0.34305
183	2	17.027	8	18.02	0.789	13.363	4.77E+01	0.01109
184	2	17.027	7.5	18.02	0.701	12.528	1.08E+02	0.0254
185	2	17.027	7	18.02	0.621	11.693	2.01E+02	0.04778
186	2	17.027	6.5	18.02	0.55	10.858	3.33E+02	0.07912
187	2	17.027	5.88	18.02	0.469	9.809	5.05E+02	0.11818
188	2	17.027	5.427	18.02	0.416	9.06	6.75E+02	0.1551
189	2	17.027	4.885	18.02	0.357	8.144	8.48E+02	0.18871
190	2	17.027	4.395	18.02	0.309	7.322	1.07E+03	0.2291
191	2	17.027	4	18.02	0.273	6.682	1.25E+03	0.25525
192	2	17.027	3.75	18.02	0.251	6.264	1.37E+03	0.27307
193	2	17.027	3.43	18.02	0.224	5.718	1.48E+03	0.2802
194	2	17.027	3.084	18.02	0.196	5.138	1.65E+03	0.29499
195	2	17.027	2.75	18.02	0.171	4.594	1.82E+03	0.30598
196	2	17.027	2.5	18.02	0.153	4.176	1.93E+03	0.30569
197	2	17.027	2.226	18.02	0.134	3.718	2.16E+03	0.3193
198	2	17.027	2	18.02	0.118	3.341	2.45E+03	0.33828
199	2	17.027	1.806	18.02	0.105	3.012	2.53E+03	0.32457
200	2	17.027	1.5	18.02	0.086	2.506	2.73E+03	0.30691
201	2	4.506	2.2	26.015	0.464	2.009	8.90E+03	0.16474
202	2	4.506	2	26.015	0.388	1.826	1.08E+04	0.20648
203	2	4.506	1.8	26.015	0.324	1.644	1.24E+04	0.24248
204	2	4.506	1.6	26.015	0.268	1.461	1.39E+04	0.2717
205	2	4.506	1.4	26.015	0.219	1.278	1.51E+04	0.29112
206	2	4.506	1.2	26.015	0.177	1.096	1.62E+04	0.29873
207	2	4.506	1	26.015	0.139	0.913	1.76E+04	0.29939
208	2	6.711	3	26.015	0.586	4.08	1.42E+03	0.08777
209	2	6.711	2.75	26.015	0.503	3.74	1.99E+03	0.12484
210	2	6.711	2.448	26.015	0.415	3.323	2.79E+03	0.17547
211	2	6.711	2.25	26.015	0.366	3.06	3.27E+03	0.20373
212	2	6.711	2.055	26.015	0.319	2.788	3.78E+03	0.23154
213	2	6.711	1.89	26.015	0.284	2.57	4.32E+03	0.26006
214	2	6.711	1.75	26.015	0.256	2.38	4.67E+03	0.27395
215	2	6.711	1.631	26.015	0.233	2.218	4.93E+03	0.28275

Table B.1: SLAC Deuteron DIS Data

I	J	E_0	E'	θ	x	Q^2	σ	F_2
216	2	6.711	1.5	26.015	0.209	2.04	5.35E+03	0.29619
217	2	6.711	1.381	26.015	0.188	1.878	5.74E+03	0.30564
218	2	6.711	1.19	26.015	0.156	1.618	6.20E+03	0.30534
219	2	6.711	1	26.015	0.127	1.36	7.56E+03	0.33379
220	2	8.713	3.751	26.015	0.711	6.623	2.39E+02	0.03213
221	2	8.713	3.502	26.015	0.632	6.183	4.20E+02	0.05681
222	2	8.713	3.252	26.015	0.56	5.742	6.41E+02	0.08669
223	2	8.713	3.001	26.015	0.494	5.299	9.12E+02	0.12228
224	2	8.713	2.751	26.015	0.434	4.857	1.21E+03	0.15949
225	2	8.713	2.448	26.015	0.367	4.315	1.61E+03	0.20483
226	2	8.713	2.272	26.015	0.332	4.011	1.71E+03	0.21237
227	2	8.713	2.139	26.015	0.306	3.777	1.89E+03	0.22922
228	2	8.713	2.002	26.015	0.281	3.535	1.95E+03	0.2298
229	2	8.713	1.88	26.015	0.259	3.319	2.43E+03	0.27865
230	2	8.713	1.753	26.015	0.237	3.095	2.56E+03	0.28473
231	2	8.713	1.501	26.015	0.196	2.65	3.11E+03	0.31908
232	2	8.713	1.252	26.015	0.158	2.211	3.50E+03	0.3236
233	2	11.9	4.499	26.015	0.781	10.849	4.14E+01	0.01408
234	2	11.9	4.249	26.015	0.714	10.246	7.42E+01	0.02508
235	2	11.9	4	26.015	0.651	9.646	1.30E+02	0.04339
236	2	11.9	3.75	26.015	0.591	9.043	1.99E+02	0.06557
237	2	11.9	3.5	26.015	0.535	8.44	2.78E+02	0.08953
238	2	11.9	3.25	26.015	0.483	7.837	3.84E+02	0.12027
239	2	11.9	3	26.015	0.433	7.234	5.02E+02	0.15205
240	2	11.9	2.75	26.015	0.386	6.631	6.21E+02	0.18068
241	2	11.9	2.5	26.015	0.342	6.029	7.78E+02	0.21564
242	2	11.9	2.197	26.015	0.29	5.281	9.16E+02	0.23598
243	2	11.9	2	26.015	0.26	4.823	1.10E+03	0.2676
244	2	11.9	1.67	26.015	0.21	4.027	1.36E+03	0.29898
245	2	15.022	5.25	26.015	0.872	15.981	5.34E+00	0.00358
246	2	15.022	5	26.015	0.809	15.22	1.03E+01	0.00684
247	2	15.022	4.75	26.015	0.75	14.459	2.27E+01	0.0149
248	2	15.022	4.5	26.015	0.694	13.698	4.10E+01	0.02643
249	2	15.022	4.25	26.015	0.64	12.937	6.48E+01	0.04096
250	2	15.022	4	26.015	0.589	12.176	9.95E+01	0.06119
251	2	15.022	3.75	26.015	0.54	11.415	1.41E+02	0.0845
252	2	15.022	3.5	26.015	0.493	10.654	1.79E+02	0.10329
253	2	15.022	3.25	26.015	0.448	9.893	2.44E+02	0.13534
254	2	15.022	3	26.015	0.405	9.132	3.06E+02	0.16269
255	2	15.022	2.75	26.015	0.364	8.371	3.78E+02	0.19032
256	2	15.022	2.439	26.015	0.314	7.407	4.70E+02	0.21906
257	2	15.022	2.25	26.015	0.286	6.849	5.31E+02	0.23546
258	2	15.022	2	26.015	0.249	6.088	6.40E+02	0.26225
259	2	15.022	1.75	26.015	0.214	5.327	7.79E+02	0.29214
260	2	18.028	5.496	26.015	0.854	20.078	3.50E+00	0.00389
261	2	18.028	5.246	26.015	0.799	19.165	7.14E+00	0.00782
262	2	18.028	4.996	26.015	0.746	18.252	1.36E+01	0.01462
263	2	18.028	4.619	26.015	0.669	16.835	2.76E+01	0.02861
264	2	18.028	4.244	26.015	0.599	15.504	5.19E+01	0.05164
265	2	18.028	3.994	26.015	0.554	14.591	7.18E+01	0.06913
266	2	18.028	3.745	26.015	0.51	13.681	9.84E+01	0.09136
267	2	18.028	3.493	26.015	0.468	12.761	1.32E+02	0.11742
268	2	18.028	3.247	26.015	0.428	11.862	1.67E+02	0.14188
269	2	18.028	3	26.015	0.389	10.96	2.05E+02	0.16544

Table B.1: SLAC Deuteron DIS Data

I	J	E_0	E'	θ	x	Q^2	σ	F_2
270	2	18.028	2.743	26.015	0.349	10.021	2.41E+02	0.18324
271	2	18.028	2.5	26.015	0.313	9.133	2.95E+02	0.20998
272	2	18.028	2.25	26.015	0.278	8.22	3.69E+02	0.24324
273	2	18.028	2	26.015	0.243	7.306	4.27E+02	0.25969
274	2	4.502	1.8	34.009	0.547	2.772	2.24E+03	0.11976
275	2	4.502	1.6	34.009	0.453	2.464	3.05E+03	0.16391
276	2	4.502	1.4	34.009	0.37	2.156	4.08E+03	0.21573
277	2	4.502	1.2	34.009	0.298	1.848	5.04E+03	0.25601
278	2	4.502	1	34.009	0.234	1.54	6.23E+03	0.29547
279	2	5.808	2.25	34.009	0.67	4.471	4.66E+02	0.05279
280	2	5.808	2	34.009	0.556	3.974	8.84E+02	0.09932
281	2	5.808	1.75	34.009	0.457	3.477	1.43E+03	0.15637
282	2	5.808	1.5	34.009	0.369	2.98	1.97E+03	0.20615
283	2	5.808	1.25	34.009	0.29	2.484	2.63E+03	0.25598
284	2	5.808	1	34.009	0.22	1.987	3.27E+03	0.28439
285	2	7.912	2.75	34.009	0.768	7.443	6.98E+01	0.01945
286	2	7.912	2.5	34.009	0.666	6.767	1.73E+02	0.04712
287	2	7.912	2.25	34.009	0.573	6.09	2.99E+02	0.07901
288	2	7.912	2	34.009	0.488	5.413	4.76E+02	0.12061
289	2	7.912	1.751	34.009	0.41	4.739	7.16E+02	0.17062
290	2	7.912	1.48	34.009	0.332	4.006	1.08E+03	0.23595
291	2	10.014	3	34.009	0.781	10.277	2.69E+01	0.01447
292	2	10.014	2.75	34.009	0.691	9.421	5.63E+01	0.02931
293	2	10.014	2.5	34.009	0.607	8.565	1.04E+02	0.05187
294	2	10.014	2.25	34.009	0.529	7.708	2.04E+02	0.09723
295	2	10.014	2	34.009	0.456	6.852	2.94E+02	0.1316
296	2	10.014	1.75	34.009	0.387	5.995	4.32E+02	0.17837
297	2	10.014	1.5	34.009	0.322	5.139	6.26E+02	0.23487
298	2	12.518	3.25	34.009	0.8	13.918	9.45E+00	0.0093
299	2	12.518	3	34.009	0.719	12.847	2.37E+01	0.02249
300	2	12.518	2.75	34.009	0.642	11.777	4.55E+01	0.04124
301	2	12.518	2.5	34.009	0.569	10.706	7.88E+01	0.06756
302	2	12.518	2.25	34.009	0.5	9.635	1.30E+02	0.10424
303	2	12.518	2	34.009	0.434	8.565	1.94E+02	0.14486
304	2	12.518	1.75	34.009	0.371	7.494	2.63E+02	0.17927
305	2	15.012	3.25	34.009	0.756	16.691	9.17E+00	0.01406
306	2	15.012	3	34.009	0.683	15.407	1.92E+01	0.02805
307	2	15.012	2.75	34.009	0.614	14.123	3.70E+01	0.0514
308	2	15.012	2.5	34.009	0.547	12.839	5.66E+01	0.0738
309	2	15.012	2.25	34.009	0.482	11.555	9.23E+01	0.11222
310	2	15.012	2	34.009	0.421	10.271	1.25E+02	0.13956
311	3	20.005	17.834	4	0.427	1.738	7.66E+05	0.18926
312	3	20.005	17.58	4	0.376	1.713	7.80E+05	0.21492
313	3	20.005	17.275	4	0.328	1.683	7.76E+05	0.24029
314	3	20.005	16.871	4	0.279	1.643	7.43E+05	0.26327
315	3	20.005	16.402	4	0.236	1.598	7.01E+05	0.28471
316	3	20.005	15.424	4	0.174	1.5	6.00E+05	0.3061
317	3	20.005	14.114	4	0.124	1.376	4.90E+05	0.31588
318	3	20.005	13.259	4	0.102	1.292	4.34E+05	0.31483
319	3	20.005	12.342	4	0.084	1.203	3.92E+05	0.31545
320	3	20.005	11.36	4	0.068	1.107	3.50E+05	0.30863
321	3	18.01	16.005	4	0.373	1.404	1.20E+06	0.22184
322	3	18.01	15.726	4	0.322	1.379	1.18E+06	0.24724
323	3	18.01	15.375	4	0.273	1.348	1.11E+06	0.26857

Table B.1: SLAC Deuteron DIS Data

I	J	E_0	E'	θ	x	Q^2	σ	F_2
324	3	18.01	14.865	4	0.221	1.303	9.98E+05	0.28684
325	3	18.01	14.174	4	0.172	1.24	8.74E+05	0.30402
326	3	18.01	13.072	4	0.124	1.147	6.96E+05	0.30688
327	3	18.01	12.276	4	0.1	1.077	6.05E+05	0.30458
328	3	18.01	11.417	4	0.081	1.002	5.32E+05	0.30117
329	3	18.01	10.494	4	0.065	0.921	4.74E+05	0.2969
330	3	16	14.143	4	0.316	1.102	1.84E+06	0.24908
331	3	16	13.838	4	0.266	1.078	1.72E+06	0.26978
332	3	16	13.435	4	0.217	1.046	1.53E+06	0.28434
333	3	16	12.81	4	0.167	0.997	1.29E+06	0.2966
334	3	16	11.992	4	0.124	0.934	1.07E+06	0.30382
335	3	16	11.208	4	0.097	0.874	8.71E+05	0.29153
336	3	16	10.408	4	0.077	0.811	7.68E+05	0.29345
337	3	12.987	11.318	4	0.228	0.716	3.45E+06	0.27537
338	3	12.987	10.947	4	0.181	0.692	2.92E+06	0.28396
339	3	12.987	10.41	4	0.136	0.658	2.30E+06	0.27988
340	3	12.987	9.768	4	0.102	0.617	1.82E+06	0.27367
341	3	12.987	9.147	4	0.08	0.579	1.50E+06	0.26434
342	4	18.001	9.606	14.991	0.747	11.768	1.42E+02	0.01789
343	4	18.001	9.33	14.991	0.702	11.428	2.05E+02	0.02624
344	4	18.001	8.972	14.991	0.649	10.989	3.21E+02	0.04186
345	4	18.001	8.646	14.991	0.603	10.59	4.22E+02	0.05587
346	4	18.001	8.244	14.991	0.551	10.095	5.82E+02	0.07807
347	4	18.001	7.794	14.991	0.498	9.542	7.74E+02	0.10505
348	4	18.001	7.339	14.991	0.449	8.984	9.89E+02	0.13494
349	4	18.001	6.908	14.991	0.406	8.457	1.15E+03	0.15733
350	4	18.001	6.487	14.991	0.368	7.942	1.37E+03	0.18531
351	4	18.001	6.102	14.991	0.335	7.469	1.57E+03	0.21028
352	4	18.001	5.832	14.991	0.313	7.144	1.64E+03	0.21843
353	4	15.004	8.86	14.991	0.785	9.047	2.19E+02	0.01473
354	4	15.004	8.625	14.991	0.736	8.807	3.27E+02	0.02261
355	4	15.004	8.346	14.991	0.682	8.52	4.92E+02	0.03494
356	4	15.004	8.046	14.991	0.629	8.214	7.13E+02	0.05192
357	4	15.004	7.723	14.991	0.577	7.884	9.79E+02	0.07308
358	4	15.004	7.369	14.991	0.525	7.521	1.28E+03	0.09731
359	4	15.004	6.984	14.991	0.474	7.128	1.61E+03	0.12487
360	4	15.004	6.627	14.991	0.43	6.761	1.92E+03	0.15096
361	4	15.004	6.191	14.991	0.382	6.317	2.29E+03	0.18218
362	4	15.004	5.895	14.991	0.352	6.018	2.44E+03	0.19458
363	4	12.497	7.705	14.991	0.729	6.553	7.76E+02	0.02887
364	4	12.497	7.495	14.991	0.679	6.374	1.07E+03	0.04123
365	4	12.497	7.245	14.991	0.625	6.16	1.51E+03	0.05983
366	4	12.497	6.952	14.991	0.568	5.911	1.98E+03	0.08168
367	4	12.497	6.635	14.991	0.513	5.64	2.61E+03	0.11106
368	4	12.497	6.292	14.991	0.459	5.348	3.18E+03	0.13976
369	4	12.497	5.981	14.991	0.416	5.084	3.66E+03	0.16406
370	4	12.497	5.733	14.991	0.384	4.877	4.07E+03	0.1852
371	4	9.999	6.4	14.991	0.645	4.355	3.40E+03	0.06203
372	4	9.999	6.215	14.991	0.596	4.229	4.23E+03	0.08025
373	4	9.999	5.997	14.991	0.543	4.08	5.11E+03	0.10099
374	4	9.999	5.824	14.991	0.506	3.963	5.94E+03	0.12097
375	4	20.001	8.562	18.996	0.869	18.647	6.58E+00	0.00309
376	4	20.001	8.307	18.996	0.824	18.091	1.22E+01	0.00575
377	4	20.001	7.978	18.996	0.77	17.373	2.40E+01	0.01133

Table B.1: SLAC Deuteron DIS Data

I	J	E_0	E'	θ	x	Q^2	σ	F_2
378	4	20.001	7.656	18.996	0.72	16.671	4.11E+01	0.01936
379	4	20.001	7.309	18.996	0.668	15.912	6.46E+01	0.03021
380	4	20.001	6.939	18.996	0.616	15.105	9.78E+01	0.04536
381	4	20.001	6.544	18.996	0.564	14.243	1.46E+02	0.06665
382	4	20.001	6.16	18.996	0.516	13.405	1.96E+02	0.08802
383	4	20.001	5.778	18.996	0.471	12.573	2.62E+02	0.11505
384	4	20.001	5.465	18.996	0.436	11.895	3.13E+02	0.13467
385	4	20.001	5.239	18.996	0.412	11.412	3.20E+02	0.13544
386	4	18.018	8.108	18.996	0.855	15.908	1.32E+01	0.00451
387	4	18.018	7.855	18.996	0.808	15.411	2.27E+01	0.00779
388	4	18.018	7.545	18.996	0.753	14.8	4.17E+01	0.01437
389	4	18.018	7.235	18.996	0.701	14.192	6.90E+01	0.02381
390	4	18.018	6.908	18.996	0.65	13.549	1.10E+02	0.0379
391	4	18.018	6.565	18.996	0.599	12.875	1.62E+02	0.05564
392	4	18.018	6.186	18.996	0.546	12.129	2.30E+02	0.0781
393	4	18.018	5.766	18.996	0.492	11.304	3.20E+02	0.10675
394	4	18.018	5.477	18.996	0.456	10.74	3.82E+02	0.12548
395	4	18.018	5.268	18.996	0.432	10.338	4.07E+02	0.1321
396	4	15.004	7.275	18.996	0.82	11.886	4.19E+01	0.00813
397	4	15.004	7.063	18.996	0.774	11.54	6.82E+01	0.01336
398	4	15.004	6.809	18.996	0.723	11.123	1.08E+02	0.02138
399	4	15.004	6.52	18.996	0.669	10.65	1.76E+02	0.03507
400	4	15.004	6.223	18.996	0.617	10.164	2.57E+02	0.0514
401	4	15.004	5.91	18.996	0.566	9.651	3.64E+02	0.07306
402	4	15.004	5.574	18.996	0.514	9.101	4.80E+02	0.09607
403	4	15.004	5.326	18.996	0.479	8.703	5.88E+02	0.11702
404	4	12.498	6.473	18.996	0.779	8.81	1.47E+02	0.01602
405	4	12.498	6.258	18.996	0.727	8.515	2.27E+02	0.02522
406	4	12.498	6.008	18.996	0.671	8.175	3.47E+02	0.03918
407	4	12.498	5.766	18.996	0.621	7.845	4.83E+02	0.05529
408	4	12.498	5.514	18.996	0.573	7.503	6.58E+02	0.07611
409	4	12.498	5.311	18.996	0.536	7.229	8.03E+02	0.09342
410	4	18.018	5.606	25.993	0.877	20.427	2.30E+00	0.00257
411	4	18.018	5.409	25.993	0.833	19.705	4.58E+00	0.00506
412	4	18.018	5.164	25.993	0.78	18.814	9.28E+00	0.01007
413	4	18.018	4.928	25.993	0.731	17.951	1.54E+01	0.01642
414	4	18.018	4.652	25.993	0.675	16.935	2.74E+01	0.02838
415	4	18.018	4.351	25.993	0.617	15.832	4.44E+01	0.04445
416	4	18.018	4.068	25.993	0.566	14.806	6.76E+01	0.06536
417	4	18.018	3.823	25.993	0.522	13.911	8.95E+01	0.08346
418	4	18.018	3.579	25.993	0.48	13.018	1.25E+02	0.11241
419	4	18.018	3.451	25.993	0.46	12.579	1.25E+02	0.10963
420	4	15.005	5.173	25.993	0.851	15.698	7.09E+00	0.00471
421	4	15.005	4.979	25.993	0.803	15.108	1.28E+01	0.00846
422	4	15.005	4.766	25.993	0.752	14.458	2.22E+01	0.01447
423	4	15.005	4.515	25.993	0.696	13.696	3.88E+01	0.02486
424	4	15.005	4.26	25.993	0.641	12.926	6.43E+01	0.04035
425	4	15.005	3.998	25.993	0.587	12.133	9.91E+01	0.06067
426	4	15.005	3.747	25.993	0.538	11.371	1.36E+02	0.08075
427	4	15.005	3.499	25.993	0.492	10.616	1.88E+02	0.1078
428	4	11.884	4.609	25.993	0.811	11.076	2.69E+01	0.0091
429	4	11.884	4.46	25.993	0.77	10.72	4.26E+01	0.01435
430	4	11.884	4.247	25.993	0.712	10.208	8.04E+01	0.02696
431	4	11.884	3.998	25.993	0.649	9.608	1.26E+02	0.0417

I	J	E_0	E'	θ	x	Q^2	σ	F_2
432	4	11.884	3.748	25.993	0.59	9.007	2.04E+02	0.06656
433	4	11.884	3.497	25.993	0.534	8.404	2.75E+02	0.08776
434	4	8.7	3.823	25.993	0.735	6.725	2.04E+02	0.02715
435	4	8.7	3.678	25.993	0.686	6.47	2.90E+02	0.03877
436	4	8.7	3.517	25.993	0.636	6.187	4.03E+02	0.05408
437	4	8.7	3.392	25.993	0.599	5.97	5.06E+02	0.0679
438	4	15.004	3.5	33.992	0.831	17.948	3.25E+00	0.00519
439	4	15.004	3.25	33.992	0.756	16.666	8.21E+00	0.01258
440	4	15.004	3	33.992	0.683	15.384	1.87E+01	0.02732
441	4	15.004	2.75	33.992	0.613	14.102	3.59E+01	0.04968
442	4	15.004	2.5	33.992	0.546	12.82	5.83E+01	0.07587
443	4	12.499	3.25	33.992	0.8	13.883	9.36E+00	0.00916
444	4	12.499	3	33.992	0.719	12.815	2.27E+01	0.02139
445	4	12.499	2.75	33.992	0.642	11.747	4.56E+01	0.04113
446	4	12.499	2.5	33.992	0.569	10.679	8.29E+01	0.07079
447	4	9.999	3	33.992	0.781	10.252	2.57E+01	0.01376
448	4	9.999	2.75	33.992	0.691	9.398	6.07E+01	0.03148
449	4	9.999	2.5	33.992	0.607	8.543	1.19E+02	0.05926
450	4	9.999	2.25	33.992	0.529	7.689	1.93E+02	0.09134
451	5	19.505	1.551	60	0.896	30.202	9.75E-02	0.00125
452	5	19.505	1.47	60	0.845	28.611	2.46E-01	0.00301
453	5	19.505	1.376	60	0.786	26.728	6.40E-01	0.00738
454	5	19.505	1.278	60	0.726	24.839	1.40E+00	0.01509
455	5	19.505	1.183	60	0.668	22.969	2.82E+00	0.02877
456	5	19.505	1.1	60	0.62	21.402	4.62E+00	0.04418
457	5	19.505	1.039	60	0.584	20.245	5.61E+00	0.05123
458	5	16.002	1.504	59.999	0.884	24.041	2.28E-01	0.00187
459	5	16.002	1.434	59.999	0.838	22.921	4.83E-01	0.00382
460	5	16.002	1.35	59.999	0.784	21.562	1.07E+00	0.00799
461	5	16.002	1.26	59.999	0.726	20.094	2.35E+00	0.01666
462	5	16.002	1.174	59.999	0.674	18.756	4.08E+00	0.02709
463	5	16.002	1.109	59.999	0.634	17.709	5.96E+00	0.03794
464	5	16.002	1.044	59.999	0.595	16.706	7.92E+00	0.04776
465	5	13.29	1.461	60	0.874	19.394	4.66E-01	0.00249
466	5	13.29	1.39	60	0.826	18.449	1.08E+00	0.00553
467	5	13.29	1.313	60	0.775	17.41	1.99E+00	0.00981
468	5	13.29	1.233	60	0.723	16.35	3.97E+00	0.01865
469	5	13.29	1.154	60	0.672	15.304	6.72E+00	0.02989
470	5	13.29	1.086	60	0.629	14.397	1.03E+01	0.04365
471	5	13.29	1.02	60	0.587	13.522	1.52E+01	0.0612
472	5	13.29	0.963	60	0.552	12.777	2.02E+01	0.07723
473	5	10.4	1.374	60	0.843	14.279	1.81E+00	0.00539
474	5	10.4	1.303	60	0.792	13.526	3.52E+00	0.01009
475	5	10.4	1.222	60	0.736	12.683	6.69E+00	0.0183
476	5	10.4	1.142	60	0.682	11.851	1.10E+01	0.02856
477	5	10.4	1.069	60	0.634	11.103	1.79E+01	0.04432
478	5	10.4	1.004	60	0.591	10.415	2.53E+01	0.05977
479	5	10.4	0.934	60	0.545	9.688	3.71E+01	0.08297
480	5	10.4	0.875	60	0.509	9.093	5.28E+01	0.11166
481	5	6.5	1.17	59.999	0.76	7.598	2.20E+01	0.02022
482	5	6.5	1.099	59.999	0.704	7.136	3.58E+01	0.03186
483	5	6.5	1.035	59.999	0.656	6.722	5.35E+01	0.04587
484	5	6.5	0.976	59.999	0.611	6.335	7.90E+01	0.06529
485	5	6.5	0.914	59.999	0.566	5.93	1.10E+02	0.08729

Table B.1: SLAC Deuteron DIS Data

I	J	E_0	E'	θ	x	Q^2	σ	F_2
486	5	6.5	0.871	59.999	0.536	5.66	1.49E+02	0.1141
487	5	19.5	1.703	49.994	0.709	23.679	2.29E+00	0.01647
488	5	19.5	1.593	49.994	0.658	22.109	4.23E+00	0.02871
489	5	19.5	1.484	49.994	0.61	20.633	7.33E+00	0.04705
490	5	19.5	1.388	49.994	0.567	19.287	1.04E+01	0.06299
491	5	19.5	1.289	49.994	0.523	17.886	1.49E+01	0.0846
492	5	19.5	1.183	49.994	0.477	16.388	2.25E+01	0.11887
493	5	19.5	1.083	49.994	0.435	15.019	2.90E+01	0.14209
494	5	16	1.709	49.995	0.727	19.508	3.72E+00	0.0176
495	5	16	1.603	49.995	0.676	18.264	5.81E+00	0.02606
496	5	16	1.48	49.995	0.62	16.885	1.01E+01	0.04277
497	5	16	1.38	49.995	0.574	15.736	1.47E+01	0.05828
498	5	16	1.275	49.995	0.526	14.522	2.37E+01	0.08814
499	5	16	1.15	49.995	0.469	13.071	3.04E+01	0.10382
500	5	16	1.059	49.995	0.432	12.103	4.64E+01	0.1488
501	5	16	0.997	49.995	0.405	11.394	5.73E+01	0.1745
502	5	16	0.933	49.995	0.377	10.663	4.83E+01	0.13958
503	5	13.5	1.698	49.998	0.738	16.349	5.00E+00	0.01628
504	5	13.5	1.608	49.998	0.694	15.487	7.16E+00	0.02232
505	5	13.5	1.513	49.998	0.647	14.563	1.28E+01	0.03819
506	5	13.5	1.407	49.998	0.596	13.527	1.76E+01	0.04956
507	5	13.5	1.303	49.998	0.548	12.54	2.77E+01	0.07337
508	5	13.5	1.207	49.998	0.503	11.612	3.94E+01	0.09804
509	5	13.5	1.122	49.998	0.464	10.78	5.20E+01	0.12169
510	5	13.5	1.042	49.998	0.429	10.028	6.35E+01	0.14002
511	5	13.5	0.983	49.998	0.403	9.464	6.87E+01	0.14535
512	5	7	1.502	49.998	0.727	7.504	3.39E+01	0.02264
513	5	7	1.425	49.998	0.681	7.119	6.09E+01	0.03957
514	5	7	1.33	49.998	0.624	6.64	9.34E+01	0.05834
515	5	7	1.233	49.998	0.569	6.155	1.29E+02	0.07728
516	5	7	1.15	49.998	0.524	5.747	1.78E+02	0.10183
517	5	7	1.075	49.998	0.483	5.366	2.37E+02	0.13064
518	5	7	1.001	49.998	0.444	4.997	2.78E+02	0.14647
519	5	7	0.917	49.998	0.401	4.575	3.63E+02	0.18064
520	5	7	0.838	49.998	0.361	4.176	4.34E+02	0.20339
521	5	7	0.772	49.998	0.33	3.861	4.37E+02	0.19349
522	6	13.5	11.36	6	0.418	1.68	3.57E+05	0.19871
523	6	13.5	11.105	6	0.365	1.642	3.56E+05	0.22101
524	6	13.5	10.862	6	0.324	1.606	3.76E+05	0.25624
525	6	13.5	9.852	6	0.213	1.457	3.22E+05	0.29728
526	6	13.5	8.494	6	0.134	1.256	2.57E+05	0.31228
527	6	13.5	6.89	6	0.082	1.019	2.10E+05	0.30991
528	6	13.5	5.997	6	0.063	0.887	1.90E+05	0.29733
529	6	16	13.554	6	0.518	2.376	1.54E+05	0.13772
530	6	16	13.236	6	0.447	2.32	1.65E+05	0.16654
531	6	16	12.998	6	0.404	2.278	1.85E+05	0.20217
532	6	16	12.585	6	0.344	2.206	1.86E+05	0.22951
533	6	16	12.006	6	0.281	2.105	1.87E+05	0.26723
534	6	16	11.367	6	0.229	1.993	1.76E+05	0.28748
535	6	16	10.664	6	0.187	1.869	1.64E+05	0.30441
536	6	16	9.904	6	0.152	1.736	1.52E+05	0.31341
537	6	16	9.081	6	0.123	1.592	1.41E+05	0.32021
538	6	16	7.256	6	0.078	1.272	1.24E+05	0.32173
539	6	19.5	16.386	6	0.599	3.5	4.96E+04	0.08376

Table B.1: SLAC Deuteron DIS Data

I	J	E_0	E'	θ	x	Q^2	σ	F_2
540	6	19.5	16.146	6	0.548	3.449	5.91E+04	0.10729
541	6	19.5	15.847	6	0.494	3.385	6.84E+04	0.13479
542	6	19.5	15.495	6	0.44	3.31	7.56E+04	0.16256
543	6	19.5	14.926	6	0.372	3.189	8.51E+04	0.20748
544	6	19.5	14.3	6	0.313	3.055	8.70E+04	0.23888
545	6	19.5	13.233	6	0.24	2.819	8.57E+04	0.27728
546	6	19.5	12.059	6	0.185	2.576	8.10E+04	0.30271
547	6	19.5	11.192	6	0.153	2.391	7.81E+04	0.31699
548	6	19.5	10.266	6	0.127	2.193	7.32E+04	0.31864
549	6	19.5	9.28	6	0.103	1.983	6.96E+04	0.31993
550	6	19.5	8.234	6	0.083	1.759	6.78E+04	0.32291
551	6	19.5	7.128	6	0.066	1.523	6.84E+04	0.32904
552	6	13.3	8.065	15	0.744	7.309	5.13E+02	0.02351
553	6	13.3	7.879	15	0.702	7.14	7.06E+02	0.03315
554	6	13.3	7.669	15	0.658	6.951	9.61E+02	0.04631
555	6	13.3	7.455	15	0.616	6.757	1.22E+03	0.06039
556	6	13.3	6.996	15	0.536	6.341	1.84E+03	0.09485
557	6	13.3	6.231	15	0.426	5.648	2.87E+03	0.1562
558	6	13.3	5.616	15	0.353	5.09	3.72E+03	0.20751
559	6	13.3	4.691	15	0.263	4.252	4.70E+03	0.26252
560	6	13.3	3.528	15	0.174	3.198	6.19E+03	0.32521
561	6	16	9.275	15	0.801	10.111	1.39E+02	0.01159
562	6	16	9.06	15	0.758	9.877	2.09E+02	0.01777
563	6	16	8.697	15	0.692	9.482	3.60E+02	0.03156
564	6	16	8.173	15	0.607	8.912	6.38E+02	0.05809
565	6	16	7.625	15	0.529	8.314	9.80E+02	0.09192
566	6	16	6.724	15	0.421	7.332	1.61E+03	0.15508
567	6	16	6.014	15	0.35	6.558	2.10E+03	0.20257
568	6	16	4.967	15	0.262	5.416	2.82E+03	0.26358
569	6	16	3.685	15	0.174	4.018	3.70E+03	0.31258
570	6	19.5	10.466	15	0.82	13.907	4.48E+01	0.00715
571	6	19.5	10.264	15	0.787	13.638	6.56E+01	0.01058
572	6	19.5	10.022	15	0.749	13.318	9.90E+01	0.01618
573	6	19.5	9.514	15	0.675	12.643	1.92E+02	0.03212
574	6	19.5	8.913	15	0.596	11.844	3.24E+02	0.05539
575	6	19.5	8.261	15	0.521	10.978	5.25E+02	0.09095
576	6	19.5	7.206	15	0.415	9.576	8.89E+02	0.1535
577	6	6.5	3.925	18	0.517	2.496	1.18E+04	0.13225
578	6	6.5	3.826	18	0.485	2.434	1.29E+04	0.14811
579	6	6.5	3.603	18	0.422	2.292	1.52E+04	0.18422
580	6	6.5	3.187	18	0.326	2.028	1.80E+04	0.23636
581	6	6.5	2.714	18	0.243	1.727	2.00E+04	0.27725
582	6	6.5	2.193	18	0.173	1.395	2.10E+04	0.29081
583	6	10.4	5.925	18	0.718	6.03	6.72E+02	0.03218
584	6	10.4	5.692	18	0.656	5.791	1.03E+03	0.05073
585	6	10.4	5.524	18	0.615	5.623	1.32E+03	0.06637
586	6	10.4	5.209	18	0.544	5.303	1.81E+03	0.09462
587	6	10.4	4.61	18	0.432	4.693	2.89E+03	0.15752
588	6	10.4	4.133	18	0.358	4.208	3.78E+03	0.20882
589	6	10.4	3.426	18	0.267	3.488	4.96E+03	0.27048
590	6	10.4	2.552	18	0.176	2.598	6.34E+03	0.32001
591	6	13.3	7.071	18	0.787	9.203	1.29E+02	0.01355
592	6	13.3	6.898	18	0.747	8.979	1.92E+02	0.02039
593	6	13.3	6.631	18	0.69	8.633	3.00E+02	0.03251

Table B.1: SLAC Deuteron DIS Data

I	J	E_0	E'	θ	x	Q^2	σ	F_2
594	6	13.3	6.22	18	0.609	8.098	5.11E+02	0.05678
595	6	13.3	5.777	18	0.533	7.521	8.30E+02	0.0938
596	6	13.3	5.047	18	0.424	6.571	1.36E+03	0.15461
597	6	13.3	4.483	18	0.353	5.836	1.83E+03	0.20593
598	6	13.3	3.666	18	0.264	4.773	2.47E+03	0.26323
599	6	13.3	2.688	18	0.176	3.5	3.32E+03	0.31086
600	6	16	7.923	18	0.819	12.407	3.97E+01	0.00749
601	6	16	7.635	18	0.762	11.954	7.69E+01	0.01472
602	6	16	7.174	18	0.678	11.236	1.62E+02	0.03146
603	6	16	6.688	18	0.599	10.475	2.90E+02	0.05695
604	6	16	6.165	18	0.523	9.656	4.62E+02	0.09077
605	6	16	5.331	18	0.417	8.349	8.03E+02	0.15482
606	6	16	4.696	18	0.347	7.355	1.09E+03	0.20231
607	6	16	3.793	18	0.259	5.941	1.54E+03	0.26316
608	6	16	2.739	18	0.172	4.29	2.12E+03	0.30726
609	6	19.5	8.941	18	0.861	17.063	9.60E+00	0.00335
610	6	19.5	8.626	18	0.807	16.462	2.10E+01	0.00739
611	6	19.5	8.228	18	0.742	15.706	4.48E+01	0.01585
612	6	19.5	7.742	18	0.67	14.778	8.65E+01	0.03062
613	6	19.5	7.174	18	0.592	13.694	1.58E+02	0.05559
614	6	19.5	6.573	18	0.517	12.547	2.56E+02	0.08881
615	6	19.5	5.628	18	0.413	10.743	4.65E+02	0.15396
616	6	19.5	4.92	18	0.343	9.391	6.29E+02	0.19687
617	6	19.5	3.93	18	0.257	7.502	9.39E+02	0.26197
618	6	19.5	2.803	18	0.171	5.35	1.36E+03	0.31063
619	6	19.5	7.662	20.6	0.86	19.104	5.55E+00	0.00333
620	6	19.5	7.363	20.6	0.806	18.353	1.18E+01	0.00703
621	6	19.5	6.977	20.6	0.74	17.398	2.50E+01	0.01483
622	7	8.005	4.199	22.238	0.7	5	7.64E+02	0.04069
623	7	8.004	4.452	13.606	0.3	2	3.35E+04	0.25124
624	7	8.004	5.34	12.418	0.4	2	4.84E+04	0.20279
625	7	9.699	4.37	19.778	0.5	5	1.86E+03	0.11827
626	7	9.692	4.847	11.843	0.22	2	3.58E+04	0.2957
627	7	12.096	3.215	20.657	0.3	5	1.76E+03	0.23763
628	7	12.094	3.896	11.826	0.13	2	1.95E+04	0.32583
629	7	12.097	5.436	15.85	0.4	5	3.42E+03	0.17643
630	7	12.097	6.768	14.196	0.5	5	3.93E+03	0.11816
631	7	12.097	7.656	13.343	0.6	5	3.53E+03	0.07286
632	7	12.097	8.29	12.82	0.7	5	2.63E+03	0.04089
633	7	15.102	3.127	11.811	0.089	2	1.22E+04	0.33247
634	7	15.102	6.221	13.247	0.3	5	4.75E+03	0.23802
635	7	15.102	7.489	17.1	0.7	10	2.15E+02	0.0279
636	7	15.102	8.441	16.101	0.8	10.001	1.18E+02	0.01125
637	7	17.308	5.197	13.54	0.22	5	3.70E+03	0.29082
638	7	17.307	6.65	16.951	0.5	10	5.21E+02	0.10235
639	7	17.307	8.426	15.047	0.6	10	4.88E+02	0.05794
640	7	21.193	7.87	14.064	0.4	9.999	9.37E+02	0.16113
641	7	21.193	7.87	17.247	0.6	14.999	1.34E+02	0.05106
642	7	24.512	5.48	11.071	0.14	5	3.67E+03	0.32716
643	7	24.503	6.74	14.135	0.3	10.001	9.31E+02	0.23158
644	8	3.748	1.084	28.728	0.2	1	1.82E+04	0.29131
645	8	4.006	1.342	24.906	0.2	1	2.47E+04	0.28724
646	8	4.251	1.586	22.205	0.2	1	3.25E+04	0.29081
647	8	5.507	2.843	14.52	0.2	1	8.71E+04	0.29246

Table B.1: SLAC Deuteron DIS Data

I	J	E_0	E'	θ	x	Q^2	σ	F_2
648	8	6.251	3.586	12.124	0.2	1	1.33E+05	0.29533
649	8	5.507	1.51	24.519	0.2	1.5	1.13E+04	0.30401
650	8	6.25	2.253	18.783	0.2	1.5	2.02E+04	0.2962
651	8	7.002	3.005	15.343	0.2	1.5	3.24E+04	0.29847
652	8	7.498	3.502	13.727	0.2	1.5	4.24E+04	0.30097
653	8	8.251	4.254	11.866	0.2	1.5	5.95E+04	0.30146
654	8	8.251	1.589	25.22	0.2	2.5	3.76E+03	0.31311
655	8	10.243	3.582	14.999	0.2	2.5	1.16E+04	0.30273
656	8	11.744	5.083	11.746	0.2	2.5	2.01E+04	0.29964
657	8	16.005	2.683	19.647	0.2	5	1.47E+03	0.29168
658	8	17.255	3.933	15.6	0.2	5	2.42E+03	0.29355
659	8	18.491	5.169	13.134	0.2	5.001	3.54E+03	0.29346
660	8	19.493	6.171	11.702	0.2	5	4.62E+03	0.29647
661	8	3.748	1.464	30.304	0.35	1.5	1.13E+04	0.23096
662	8	4.007	1.723	26.95	0.35	1.5	1.47E+04	0.23041
663	8	4.25	1.966	24.459	0.35	1.5	1.82E+04	0.22922
664	8	5.507	3.223	16.715	0.35	1.5	4.40E+04	0.23232
665	8	7.002	4.718	12.232	0.35	1.5	8.93E+04	0.23249
666	8	5.501	1.695	30.008	0.35	2.5	3.79E+03	0.2243
667	8	6.25	2.443	23.345	0.35	2.5	6.54E+03	0.22132
668	8	7.081	3.274	18.9	0.35	2.5	1.05E+04	0.22055
669	8	7.498	3.692	17.283	0.35	2.5	1.31E+04	0.2231
670	8	9.71	5.904	11.986	0.35	2.5	2.98E+04	0.22028
671	8	10.243	2.63	24.878	0.35	5	1.25E+03	0.20817
672	8	11.753	4.14	18.447	0.35	5	2.40E+03	0.20729
673	8	13.32	5.707	14.735	0.35	5	4.01E+03	0.20917
674	8	15.004	7.391	12.189	0.35	5	6.13E+03	0.20747
675	8	3.749	1.084	46.177	0.5	2.5	1.51E+03	0.13995
676	8	4.251	1.587	35.447	0.5	2.501	2.66E+03	0.14095
677	8	5.502	2.838	23.082	0.5	2.5	6.87E+03	0.14082
678	8	7.082	4.418	16.25	0.5	2.5	1.53E+04	0.14209
679	8	9.248	6.584	11.63	0.5	2.5	3.18E+04	0.13956
680	8	7.084	1.755	36.976	0.5	5.001	4.73E+02	0.11734
681	8	8.25	2.921	26.331	0.5	5.001	9.82E+02	0.11796
682	8	9.71	4.381	19.742	0.5	5.001	1.87E+03	0.11868
683	8	13.316	7.987	12.448	0.5	5	5.24E+03	0.11696
684	8	10.243	2.249	33.152	0.5	7.5	2.34E+02	0.10687
685	8	14.991	6.997	15.367	0.5	7.5	1.28E+03	0.10928
686	8	13.319	2.661	30.802	0.5	9.999	1.46E+02	0.10344
687	8	15.005	4.348	22.578	0.5	10.001	2.79E+02	0.10272
688	8	18.49	7.832	15.1	0.5	10	6.76E+02	0.10241
689	8	13.32	5.5	18.062	0.492	7.22	9.99E+02	0.11552
690	8	13.32	6.318	14.927	0.432	5.68	2.85E+03	0.15229
691	8	16.006	5.331	18.042	0.419	8.391	7.87E+02	0.15306

Appendix C
HERA Proton DIS Data

This appendix contains the F_2 values extracted from the analysis of the HERA deep inelastic scattering data for the proton collected during 1993. The data is from Table 4 of the report written by the H1 Collaboration entitled:

A Measurement of the
Proton Structure Function $F_2(x, Q^2)$,

document number DESY 95-006, dated January 1995.

The data presented here also contains modified x and F_2 data adjusted to the specifications in the discussion in Chapter 4 of this document. Specifically, the original x values are multiplied by 9 to convert them from proton-mass values to muon-mass values. In addition, 0.346 is subtracted from the original F_2 values to normalize them for the muon.

In the table:

x	=	proton momentum fraction
F_2	=	proton structure function
$x \cdot 9$	=	muon momentum fraction
$F_2 - 0.346$	=	muon structure function
Q^2	=	4-momentum transfer squared (GeV²)

Details concerning the collection, analysis, preparation and error treatment of this data are in the reference given above.

Table C.1: HERA Proton DIS Data

x	F_2	$x \cdot 9$	$F_2 - 0.346$	Q^2
0.000178	1.16	0.001602	0.814	4.5
0.000178	1.21	0.001602	0.864	6
0.000178	1.19	0.001602	0.844	8.5
0.000261	1.2	0.002349	0.854	8.5
0.000261	1.35	0.002349	1.004	12
0.000383	1.11	0.003447	0.764	8.5
0.000383	1.26	0.003447	0.914	12
0.000383	1.4	0.003447	1.054	15
0.000562	1.19	0.005058	0.844	12
0.000562	1.35	0.005058	1.004	15
0.000562	1.52	0.005058	1.174	20
0.00075	0.94	0.00675	0.594	6
0.00075	0.76	0.00675	0.414	8.5
0.000825	1.08	0.007425	0.734	12
0.000825	1.17	0.007425	0.824	15
0.000825	1.17	0.007425	0.824	20
0.000825	1.47	0.007425	1.124	25
0.000825	1.71	0.007425	1.364	35
0.00133	0.96	0.01197	0.614	12
0.00133	1.13	0.01197	0.784	15
0.00133	1.03	0.01197	0.684	20
0.00133	1.23	0.01197	0.884	25
0.00133	1.23	0.01197	0.884	35
0.00133	1.46	0.01197	1.114	50
0.00237	0.85	0.02133	0.504	12
0.00237	0.94	0.02133	0.594	15
0.00237	1.03	0.02133	0.684	20
0.00237	1.02	0.02133	0.674	25
0.00237	1.1	0.02133	0.754	35
0.00237	1.08	0.02133	0.734	50
0.00237	1.4	0.02133	1.054	65
0.00237	1.09	0.02133	0.744	80
0.00237	1.6	0.02133	1.254	120
0.00421	0.74	0.03789	0.394	12
0.00421	0.78	0.03789	0.434	15
0.00421	0.83	0.03789	0.484	20
0.00421	0.91	0.03789	0.564	25
0.00421	0.97	0.03789	0.624	35
0.00421	1	0.03789	0.654	50
0.00421	1.09	0.03789	0.744	65
0.00421	1.19	0.03789	0.844	80
0.00421	0.99	0.03789	0.644	120
0.00421	1.41	0.03789	1.064	200
0.0075	0.7	0.0675	0.354	12
0.0075	0.71	0.0675	0.364	15
0.0075	0.74	0.0675	0.394	20
0.0075	0.73	0.0675	0.384	25
0.0075	0.88	0.0675	0.534	35
0.0075	0.65	0.0675	0.304	50
0.0075	0.95	0.0675	0.604	65
0.0075	0.7	0.0675	0.354	80
0.0075	0.83	0.0675	0.484	120
0.0075	0.91	0.0675	0.564	200

Table C.1: HERA Proton DIS Data

x	F_2	$x \cdot 9$	$F_2 - 0.346$	Q^2
0.0075	1.16	0.0675	0.814	400
0.0133	0.58	0.1197	0.234	12
0.0133	0.59	0.1197	0.244	15
0.0133	0.64	0.1197	0.294	20
0.0133	0.71	0.1197	0.364	25
0.0133	0.86	0.1197	0.514	35
0.0133	0.66	0.1197	0.314	50
0.0133	0.69	0.1197	0.344	65
0.0133	0.71	0.1197	0.364	80
0.0133	0.73	0.1197	0.384	120
0.0133	0.72	0.1197	0.374	200
0.0133	0.81	0.1197	0.464	400
0.0133	1.13	0.1197	0.784	800
0.0237	0.51	0.2133	0.164	20
0.0237	0.52	0.2133	0.174	25
0.0237	0.57	0.2133	0.224	35
0.0237	0.52	0.2133	0.174	50
0.0237	0.5	0.2133	0.154	65
0.0237	0.6	0.2133	0.254	80
0.0237	0.65	0.2133	0.304	120
0.0237	0.54	0.2133	0.194	200
0.0237	0.71	0.2133	0.364	400
0.0237	0.82	0.2133	0.474	800
0.0421	0.55	0.3789	0.204	35
0.0421	0.48	0.3789	0.134	65
0.0421	0.47	0.3789	0.124	80
0.0421	0.34	0.3789	-0.006	120
0.0421	0.37	0.3789	0.024	200
0.0421	0.78	0.3789	0.434	400
0.0421	0.67	0.3789	0.324	800
0.0421	0.86	0.3789	0.514	1600
0.0422	0.4	0.3798	0.054	50
0.075	0.45	0.675	0.104	80
0.075	0.48	0.675	0.134	120
0.075	0.42	0.675	0.074	400
0.075	0.57	0.675	0.224	800
0.075	0.62	0.675	0.274	1600
0.133	0.31	1.197	-0.036	400
0.133	0.3	1.197	-0.046	800
0.133	0.37	1.197	0.024	1600

Notes

Chapter 1

[1] Ernest Rutherford predicted the existence of a particle like the neutron in 1920, 12 years before its discovery. In a 1920 lecture, Rutherford proposed:

"Under some conditions, however, it may be possible for an electron to combine much more closely with the H nucleus, forming a kind of neutral doublet. Such an atom would have very novel properties. Its external field would be practically zero, except very close to the nucleus, and in consequence it should be able to move freely through matter. Its presence would probably be difficult to detect by the spectroscope, and it may be impossible to contain it in a sealed vessel. On the other hand, it should enter readily the structure of atoms, and may either unite with the nucleus or be disintegrated by its intense field, resulting possibly in the escape of a charged H atom or an electron or both."

From: E. Rutherford, Proc. Roy. Soc., A97, 374 (1920): Bakerian Lecture: Nuclear Constitution of Atoms.

[2] The isotope decay information comes from the reference Table of the Isotopes, compiled by Norman E. Holden for Brookhaven National Laboratory (Revised 2002).

[3] The mass of a free proton is 1.007277 u, which is less than the mass of a neutron, which is 1.008665 u.

[4] Although there are a number of articles on the internet that talk about the possibility of the proton being a neutron plus a positron, I could find no formal reference to any one of the early physicists proposing it.

[5] The isotope decay information comes from the source given in note 2 above. Some references list the decay modes of some of these isotopes as electron capture.

[6] The isotope decay information comes from the source given in note 2 above.

Chapter 2

[7] A good description of what went on during these experiments is in the lecture Henry Kendall gave while receiving his 1990 Nobel prize in physics. One reference to the lecture is:

Kendall, H. W. 1990. "Deep Inelastic Scattering: Experiments on the Proton and the Observation of Scaling," Nobel Lectures in Physics 1981-1990 (Nobel Lectures, Including Presentation Speeches and Laureate), 1991, World Scientific Publishing Company, Singapore.

At the time this book was published, the above reference could be found on the internet at: URL: http://nobelprize.org/nobel_prizes/physics/laureates/1990/kendall-lecture.pdf.

[8] This is a reference to the experiments done by Rutherford and Geiger, where they fired high-energy alpha particles at various metal targets, including gold. They expected the alphas to pass through virtually uninterrupted, but the alphas scattered through large angles, with some even scattering back toward the source. It was from these experiments that Rutherford proposed that the atom is made of a positively charged core, containing most of the mass of the atom. The reference to the paper Rutherford wrote discussing his finding is:

The Scattering of α and β Particles by Matter and the Structure of the Atom. Philosophical Magazine. 1911 May; 6 (21).

[9] Chambers, E.E. and Hofstadter, R. Phys. Rev. 103, 1454-1463 (1956).

[10] A description of various types of electron scattering including deep inelastic scattering is in a report entitled: The Parton Model, by P. Hansson, dated November 18, 2004. At the time this book was published, it was at URL: http://www-zeus.desy.de/~liuc/physics/parton.pdf.

[11] A discussion of these variables and the role they play in deep inelastic scattering is in Appendix A of the Ph.D. thesis for Stanford University, written by L.W. Whitlow entitled: Deep Inelastic Structure Functions from Electron Scattering on Hydrogen, Deuterium, and Iron at $0.6 \text{ GeV}^2 < Q^2 < 30.0 \text{ GeV}^2$, document number SLAC-357, dated March 1990.

[12] This example is from experiment number 315 in the list of SLAC proton scattering experiment results listed in Appendix A.

[13] Bjorken, J.D. Phys. Rev. 179, 1547 (1969).

[14] The two plots that originally confirmed Bjorken scaling are Figure 13 and Figure 16 of the Kendall Nobel lecture reference in note 7 above.

[15] Callan C. and Gross D.J. Phys. Rev. Lett. 21, 311 (1968).

[16] The plot that originally confirmed that the scattering data satisfied the Callan-Gross relation is Figure 18 of the Kendall Nobel lecture reference in note 7 above.

Chapter 3

[17] The data for this plot is in Appendix A of this book. The fit is a fourth order polynomial calculated using the Trendline option of the chart function in Excel 2003.

[18] The muon is a particle with a charge of -1 and a spin of ½. Its mass is 105.658 MeV, which is 206.768 electron masses. Its has an antiparticle, the antimuon that has the same mass and spin, but a charge of +1. A free muon decays into an electron, an electron antineutrino and a muon neutrino. Its mean lifetime is 2.2 microseconds.

[19] The data for this plot is in Appendix B of this book. The fit is a fourth order polynomial calculated using the Trendline option of the chart function in Excel 2003.

---

[20] Leissner, Boris, Muon Pair Production in Electron-Proton Collisions, October 29, 2002.

[21] Muon Pair Production in ep Collisions at HERA, DESY-03-159, October 2003.

[22] Beatty, J.J. and J. Matthews, Cosmic Rays, August 2011.

Chapter 4

[23] Taken from the introduction of lecture: HERA the new frontier, by J Feltesse, presented at the SLAC Summer Institute, Stanford, CA, August 5 – August 16, 1991

[24] The data used to produce this table is in Appendix C of this document. The F_2 values for all Q^2 values of an x value were added, and the sum divided by the number of values. The values for x = 0.00075 were not included in the table because they were significantly outside the range of their neighboring points.

[25] To produce the F_2 values in this table, the adjustment factor of 0.346 was applied to the raw F_2 data given in Appendix C before the x values for various Q^2 values were averaged. This produced the differences between the corresponding values in Table 4.1 and Table 4.2 of exactly 0.346.

[26] This is a plot of the data from Table 4.2. The data is broken into two sets. One set contains points from x = 0.001602 to x = 0.005058. The other, points from x = 0.005058 to x = 0.67500. A logarithmic curve fit was generated for each set using the Trendline option of the chart function in Excel 2003.

[27] This is a portion of the data from Figure 4.1 blown up to show how the data behaved around the peak F_2. The plot in Figure 4.2 shows the points from x = 0.001602 to x = 0.007425.

Chapter 5

[28] The raw data for this plot is in Appendix B of this book. The F_2 values from Appendix B are multiplied times 2 to produce the F_2 values used in the plot. This was done to adjust the F_2 values to the actual values measured for the deuteron. The fit of the data is a fourth order polynomial calculated using the Trendline option of the chart function in Excel 2003.

[29] The actual deuteron F_2 curve is the bottom graph in Figure 5 of the lecture Jerome Friedman gave while receiving his 1990 Nobel prize in physics. One reference to the lecture is:

Friedman, J. I., 1990. "Deep Inelastic Scattering: Comparison With The Quark Model," Nobel Lectures in Physics 1981-1990 (Nobel Lectures, Including Presentation Speeches and Laureate), 1991, World Scientific Publishing Company, Singapore.

At the time this book was published, the above reference could be found on the internet at: URL: http://nobelprize.org/nobel_prizes/physics/laureates/1990/friedman-lecture.pdf

---

[30] This is equivalent to using the deuteron mass in equation (5.1) to calculate the momentum fraction instead of the proton mass.

[31] The raw data for this plot is in Appendix B of this book. For the plot, the momentum fraction data from Appendix B have been modified by a factor of the ratio of the proton mass to the deuteron mass (½), to make the momentum fractions plotted deuteron momentum fractions. The fit of the data is a fourth order polynomial calculated using the Trendline option of the chart function in Excel 2003.

[32] The raw data for this plot is in Appendix B of this book. The structure function values in the plot are the ones from the data multiplied by the ratio of the proton's mass to that of the deuteron. The fit of the data is a fourth order polynomial calculated using the Trendline option of the chart function in Excel 2003.

[33] The data for this plot is in Appendix B of this book. The structure function values and momentum fraction values in the plot are the ones from the data multiplied by the ratio of the proton's mass to that of the trimuon. The fit of the data is a fourth order polynomial calculated using the Trendline option of the chart function in Excel 2003.

Chapter 6

[34] I have no real basis for this claim other than the particles appear to coexist somehow within the muon. Some attractive force apparently holds the like-charged particles together in the muon, yet some repulsive force holds the electrons and positrons apart to avoid annihilation. It would seem that an arrangement of equally spaced electrons and positrons could achieve such a configuration.

[35] There is no evidence to suggest that there are any differences between the muon-muon, antimuon-antimuon, or muon-antimuon bonds. Even though the two particles have different net charges, each having an internal structure of alternating electrons and positrons should make them indistinguishable to each other at the periphery.

[36] Based on the physical appearance of the model, the proton can twist itself into many different configurations. The two shown in the figure are the ones I think most likely to occur. I have no bases for this other than a "gut" feeling. Not very scientific.

[37] Our model of the proton will be 1,837 electrons. The muons only share 26 electrons in a free proton.

[38] This model is surprisingly close to the early belief held by physicists, including Ernest Rutherford, that the neutron is just a proton and an electron.

Chapter 7

[39] I predicted the existence of single and double bonds between nucleons in my book Nuclear Alternative. There, I made models of atomic nuclei using protons and either single or double

bonds similar to the bonds described here. I had no real basis for their existence except that they worked well in the models. The model masses compared to within 3 or 4 electron masses of the actual masses of the nuclei.

[40] It is hard to say what actually happens when the bond between the proton and the neutron forms. It may be that only deuterons that result from collisions that involve the muon in the neutron with the extra electrons are stable and survive. The deuterons that form from collisions not involving the muon with the extra electrons may instantly decay. Or, perhaps when the collision occurs, the neutron relays the extra electrons through its muons and antimuons to the bonding muon. Or, it could happen some other way. All that is clear is that somehow the extra electrons get incorporated into the bond.

[41] Regardless of how the bond forms, the outcome is that the deuteron looks like two protons sharing a positron that act as the valence positron for both of them.

[42] It is generally easier to make models of nuclei out of just protons instead of protons and neutrons. It seems that all the nucleons in a nucleus are trying to acquire the number of positrons and electrons they need to "feel" like free protons. Consequently, all of the nucleons in a nucleus are protons. The illusion of neutrons arises because some protons in the nucleus share their valence positrons with other protons. The abundance of hydrogen gas in the universe makes nuclei made from only protons seem very plausible. Traditionally, neutrons only seem to appear because of a catastrophic event such as fission or particle bombardment that literally tears nucleons off nuclei.

[43] The mass dilation may be due to a significant portion of the mass of a particle going into generating the fields that exist between it and the particles around it.

[44] In electron capture, an electron from the atom's orbital cloud is pulled into the nucleus, reducing its charge by 1. It has the same effect on the nucleus as positron emission.

[45] When a helion (He-3) absorbs a neutron, instead of becoming an alpha particle (He-4), it emits a proton and becomes a triton (H-3). Based on our models of the neutron, the proton, the helion and the triton; it seems likely that the helion never actually absorbs the neutron. Instead, when the helion and the neutron collide, the neutron sets off an annihilation of the extra electron-positron pair within it to propel its remaining extra electron into the helion. In other words, the collision sets off beta decay within the neutron that causes it to inject the helion with an electron. This converts the neutron into the proton observed during the reaction, and converts the helion into the triton.

[46] As mentioned in the previous note, when a helion absorbs a neutron, it does not become an alpha particle, even though, intuitively, one might expect it to. The helion has a very high neutron absorption cross section, giving the appearance that it has a strong tendency to go to the more stable alpha particle configuration. However, it seems, as our models suggest, that it just cannot pull off the transition. Too much has to happen after the absorption for the new collection of particles to become an alpha particle. This outcome is consistent with the expected behavior of our models of the helion, the neutron and the alpha particle.

[47] Even at high energies, the probabilities of getting deuterons (D) and tritons (T) to fuse into alpha particles are very small. This is further evidence that a wholesale reconfiguration of the bonding in the colliding particles must occur for an alpha particle to form.

[48] The apparent absence of antimatter in the visible universe is an ongoing puzzle in the realm of physics. Matter born from energy usually comes with an equal amount of antimatter. However, because electrons, protons and neutrons are all considered matter, there does not appear to be any antimatter forming the materials we see in the universe. Our model of the proton (and the muon and neutron) may finally reveal the antimatter that has been hiding from us since our awareness of its necessary existence.

Index

ABOUT THE AUTHOR

William Stubbs is a retired engineer who independently researches a variety of subjects including physics. He earned a degree in Nuclear Engineering from the University of Tennessee and worked for both private and public engineering organizations during his 30-year career. His former employers include General Electric, Westinghouse Electric, the Tennessee Valley Authority, and the U.S. Department of Energy. He has published several articles on physics and nuclear science since retiring in 2005, and published two books: *Nuclear Alternative: Redesigning Our Model of the Structure of Matter* in 2008; and *Gravity* in 2012. In addition to his research, William enjoys listening to, performing, and composing music; developing simple Android apps; watching classis sci-fi movies, and spending time with family and friends. He lives in Port St. Lucie, Florida.

www.ingramcontent.com/pod-product-compliance
Lightning Source LLC
Chambersburg PA
CBHW070824180526
45168CB00002B/735